高等院校土建学科双语教材（中英文对照）
◆ 土木工程专业 ◆
BASICS

屋顶构造
ROOF CONSTRUCTION

［德］塔尼娅·布罗屈克　编著
　　杨　璐　柳美玉　译

中国建筑工业出版社

著作权合同登记图字：01-2007-3338号

图书在版编目（CIP）数据

屋顶构造/（德）布罗屈克编著；杨璐，柳美玉译. —北京：中国建筑工业出版社，2011

高等院校土建学科双语教材（中英文对照）◆ 土木工程专业 ◆
ISBN 978-7-112-12283-7

Ⅰ. 屋… Ⅱ. ①布…②杨…③柳… Ⅲ. 屋顶-建筑构造 Ⅳ. TU231

中国版本图书馆CIP数据核字（2010）第141303号

Basics：Roof Construction / Tanja Brotrück
Copyright © 2007 Birkhäuser Verlag AG（Verlag für Architektur），P. O. Box 133, 4010 Basel, Switzerland
Chinese Translation Copyright © 2011 China Architecture & Building Press
All rights reserved.
本书经Birkhäuser Verlag AG出版社授权我社翻译出版

责任编辑：孙　炼
责任设计：陈　旭
责任校对：王　颖　陈晶晶

高等院校土建学科双语教材（中英文对照）
◆ 土木工程专业 ◆
屋 顶 构 造
［德］塔尼娅·布罗屈克　编著
　　杨　璐　柳美玉　　译
*
中国建筑工业出版社出版、发行（北京西郊百万庄）
各地新华书店、建筑书店经销
北京嘉泰利德公司制版
北京云浩印刷有限责任公司印刷
*
开本：880×1230毫米　1/32　印张：4½　字数：144千字
2011年5月第一版　2011年5月第一次印刷
定价：**16.00**元
ISBN 978-7-112-12283-7
（20281）

版权所有　翻印必究
如有印装质量问题，可寄本社退换
（邮政编码100037）

中文部分目录

\\ 序　7

\\ 导言　78
　　\\ 荷载与力　79

\\ 坡屋顶　83
　　\\ 基础知识　83
　　\\ 屋顶类型　84
　　\\ 屋顶窗　87
　　\\ 屋顶构造　89
　　　　\\ 对椽屋架　89
　　　　\\ 三角屋架　92
　　　　\\ 有檩屋架　93
　　　　\\ 承接梁　96
　　　　\\ 天窗与屋顶窗　96
　　\\ 屋顶结构分层　99
　　　　\\ 屋面覆层　99
　　　　\\ 屋面木条　102
　　　　\\ 屋面防水　105
　　　　\\ 保温隔热层　107
　　\\ 面层铺设方法　109
　　\\ 屋面排水　113
　　\\ 表现方法　116

\\ 平屋顶　125
　　\\ 基础知识　125
　　\\ 屋面结构层　125
　　\\ 面层铺设方法　130
　　\\ 防水板　133
　　\\ 屋面收边　134
　　\\ 屋面排水　137

\\ 结语　141

\\ 附录 142
　　\\ 相关规范 142
　　\\ 参考文献 144
　　\\ 图片来源 144

CONTENTS

\\Foreword _9

\\Introduction _10
    \\Loads and forces _11

\\Pitched roofs _15
    \\Basics _15
    \\Roof types _16
    \\Dormers _19
    \\Roof structures _21
        \\Couple roof _21
        \\Collar roof _24
        \\Purlin roof _25
        \\Trimming _28
        \\Roof windows and dormers _28
    \\Layers of structural elements _31
        \\Roof coverings _31
        \\Roof battens _34
        \\Waterproofing _37
        \\Insulation _39
    \\Types of finish _41
    \\Drainage _45
    \\Presentation _48

\\Flat roofs _57
    \\Basics _57
    \\Layers of structural elements _57
    \\Types of finish _62
    \\Flashing _65
    \\Roof edging _66
    \\Drainage _69

\\In conclusion _73

\\Appendix _75
    \\Standards _75
    \\Literature _77
    \\Picture credits _77

序

位于我们头顶的屋顶结构可以满足人类的一个基本需求——帮助我们抵抗雨雪风霜以及寒冷的侵袭。除此之外，屋顶结构还必须能够起到传递荷载和保持稳定性的作用：屋顶结构需要具有多种不同的功能。在人类历史发展的过程中，出现了多种不同形状和类型的屋顶结构，以通过不同的建造方法来满足屋顶结构的功能需要，而到今天，我们仍然在沿袭这些方法。

另外，屋顶结构也必须满足美观的要求。屋顶常被称作为建筑的"第五立面"。建筑作为人造景观，其平屋顶和坡屋顶的不同变化形式决定了建筑的主要特征，同时也为新的建筑设计提供了重要的素材。

本套丛书的出版主要针对的是初次接触其中某一主题或者学科的学生，希望该丛书能够起到一个普及教育和实例分析的作用。丛书的内容简单易懂，而且包括了相应的实例。丛书的每一册都对一些非常重要的概念进行了详细和深入的解释。丛书并不打算对广阔的专业知识进行一个纲要性的介绍，而是旨在为读者就某一主题进行入门介绍并让读者掌握一些必要的专业知识。

本书主要是针对最初接触相关知识的未来建筑师、结构工程师以及其他建筑专业人士而出版的。该册主要对不同的屋顶类型进行了介绍，并且阐述了不同的屋顶建造方法如何满足相应的结构功能要求以及各自的优缺点。本书对屋顶结构的各个要素和建筑层进行了清晰的说明，并对如何在设计阶段对各个因素进行考虑进行了指引；还对屋顶的结构形式、保温隔热层、防水层、覆层、面层以及排水系统的基本构件进行了介绍，目的在于让初学者熟悉和了解一些必要的专业术语，帮助他们在实际的设计和建造过程中进行更好地理解和区分。

编者：Bert Bielefeld

FOREWORD

The roof over our heads satisfies a fundamental human need – it protects us from rain, wind and cold. In addition to these technical requirements it must transfer loads and provide stability: a roof has a variety of functions to fulfil. Craft traditions have generated numerous roof shapes and typologies to address these tasks in a number of ways, which are still used today.

The roof must be aesthetically satisfying as well; it is often called the fifth façade. Variants on flat and pitched roof forms define the character of whole man-made landscapes, and also offer essential stylistic resources for new buildings.

The "Basics" series of books aims to provide instructive and practical explanations for students who are approaching a subject or discipline for the very first time. It presents content with easily comprehensible introductions and examples. The most important principles are systematically elaborated and treated in depth in each volume. Instead of compiling an extensive compendium of specialist knowledge, the series aims to provide an initial introduction to a subject and give readers the necessary expertise for skilled implementation.

The "Roof" volume is aimed at students who are encountering roofs for the first time as part of their training as architects, structural engineers, or other construction professionals. It explains roof types, how construction methods meet structural requirements, and their various advantages and disadvantages. The book gives a clear account of the individual structural elements and layers, and provides guidance on addressing them at the planning stage. It deals with the essential roof structure, insulation and waterproofing, coverings and surfaces, and the basic elements of drainage. The aim is to familiarize students with the necessary technical terms, so that they can translate general facts and differences into concrete design and construction.

Bert Bielefeld, Editor

INTRODUCTION

The roof is part of a building's outer skin, and fulfils a range of functions: first, it protects the space below it, open or closed, from the weather. Here the most important aspects are draining precipitation effectively, providing protection from sun and wind, and affording privacy.

Different structures can be used according to functional requirements or the design approach. The roofs described in this book demonstrate basic principles. They form a basis for new roof planning approaches, which are in a constant state of development.

Various forces act on the roof. They must be conducted to the ground directly, or via outside walls, columns or foundations.

We distinguish between various structures and roof forms. A number of factors are involved in choosing a suitable roof. Appearance is probably the most important criterion. Then come the configuration and size of the plan view; construction costs and relevant building regulations play a crucial role.

The choice of structure and materials should be appropriate to the project in hand: elaborate prefabricated steel constructions are rarely used for private houses, and hand-finished on-site detailing is avoided for industrial buildings where possible.

Typically regional roof forms often emerge. Alpine regions usually have shallow-pitched roofs with very large overhangs, while houses with

> \\Hint:
> Roof pitch and roof shapes are often stipulated for building plots subject to a new master plan. If the plot is in a developed area and there is no master plan, "fitting in with the surroundings" is the correct approach to building regulations. The local building department will provide information about whether a particular site is subject to precise stipulations.

steeply pitched roofs set gable-on to the street › see chapter Roof types are more usual in northern European coastal regions. But buildings' functions have also produced typical roof shapes. For example, indoor tennis courts have vaulted barrel roofs that follow the flight of the ball, while normal events halls have flat roofs to facilitate flexible use.

Different roof types can be combined, but this often produces a complicated geometry of details. Simple structures are therefore preferable, to avoid leakage.

The main distinction in roof types is between pitched and flat roofs; generally speaking a roof is considered pitched if it inclines by more than 5°. These two roof forms are clearly distinct in structure and function, and will be considered separately in this book.

LOADS AND FORCES

The statics of a building deal with its structural stability: the forces acting on it and their effects have to be calculated. Newton's law says: force = mass × acceleration. As a rule, forces cannot be identified directly, but only indirectly, by their effects. For example, if we observe the acceleration of a body, we will establish that one or more forces are at work. But in building, statics is the theory of the equilibrium of forces: the various parts of the buildings should be at rest. It is also essential to ensure that the internal forces are also in equilibrium, which means that each component part has to withstand load. Its ability to do this depends on its thickness or dimensions, and on the solidity and elasticity of the material.

If a load compresses a construction element, <u>compressive forces</u> are generated. If the forces affecting the element are pulling it apart, <u>tensile forces</u> are generated. If opposing forces affect an element at different points, the element tries to twist. The building industry applies the technical term <u>momentum</u> or <u>torque</u> to this torsion. The sum of the maximum forces that could be exerted identifies the overall forces that have to be directed into the construction below and absorbed by it.

The forces affecting a building or a construction element are also defined according to their direction. A distinction is made between longitudinal forces and lateral forces.

Various forces act on buildings. They must be identified at the planning stage and plans must be made for transferring them into the

Table 1:
Loads

Type of load		Duration	Main direction	Determination
Dead load	△ ↓ ↓	Permanent	Vertical	Calculated according to the quantity and specific weights of the structural elements (in KN/m²)
Imposed load	△ ↓↓	Variable	Vertical	Can be taken from table values as a mean values for certain uses (in KN/m²)
Snow and ice load	↓ △ ↓ ↓	Variable	Vertical	Can be taken from table values according to the roof pitch and snow-loaded areas
Wind load	↘ △ ↙	Variable	Variable	Can be taken from table values according to the roof pitch and wind-loaded areas

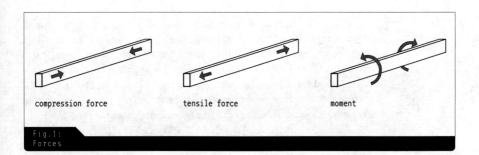

compression force   tensile force   moment

Fig.1:
Forces

ground. Loads can act horizontally, in longitudinal and transverse directions, and vertically. Identifying the individual loads forms the basis for dimensioning the roof construction. Planners must first decide which materials to use, so that the building's self-weight can be determined. The <u>dead load</u> is a permanent load. It acts vertically downwards. <u>Imposed loads</u> are the next factor. These can be movable objects, such as furniture, or people. But it is not necessary to list every object individually and take it into account when dimensioning the structure. Mean values are available for different types, e.g. dwellings, factories and warehouses. Individual specifications are required only in exceptional cases. If a structural element is not planned to be generally accessible, a diagonal roof section, for example, it is still necessary to ensure that a person could walk on it for maintenance purposes, or during the assembly process. This is known as a point load. As a rule, imposed loads act vertically downwards, like a dead load.

<u>Wind</u>, <u>snow</u> and <u>ice</u> loads act on the roof from the outside. Snow and ice exert pressure on the roof because of their weight, and so also create vertical forces, but wind can act both horizontally and vertically. These forces are identified as wind suction and wind pressure. Wind suction acts as a lifting force. Structural elements that are so loaded must be appropriately protected against being blown away.

> \\Tip:
> Individual national standards provide load compilation tables. The individual weights of materials and structural elements and assumptions about imposed, snow and wind loads can be taken from these. The most important standards are listed in the appendix to this book.

PITCHED ROOFS

BASICS

By far the most roofs for detached dwellings are pitched. Pitched roofs are exceptionally well suited to draining precipitation off buildings. The loadbearing structure is usually of wood and is made by hand, although steel and concrete are also possible. The triangular cross sections under the roof surfaces absorb horizontal wind forces well and conduct them into the structure.

The highest point of the roof is known as the <u>ridge</u>, and the lower edges as the <u>eaves</u>. The diagonal link on the wall of the house, at the <u>gable</u>, forms the <u>verge.</u> ⟩ see Fig. 2 When two roof surfaces intersect, the intersection line pointing outwards is known as the <u>arris</u> and the internal line as the <u>valley</u>. It the roof is set on a wall that rises higher than the topmost ceiling in the house, this wall is called a <u>jamb wall</u>. The <u>roof pitch</u> is defined by the angle between the roof surface and the horizontal. This dimension is always given as the inside angle and is measured in degrees. For gutters and waterproofing elements the term slope is used. This is usually given as a percentage.

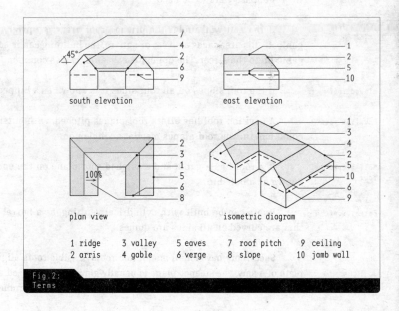

Fig. 2: Terms

1 ridge   3 valley   5 eaves   7 roof pitch   9 ceiling
2 arris   4 gable   6 verge   8 slope   10 jamb wall

Fig.3:
Monopitch roof – Gable roof – Mansard roof – Hipped roof

ROOF TYPES

The different roof forms have names that define the nature of the roof and gable pitch.

Monopitch roof — A single inclined area is called a monopitch roof. This form produces walls of different height at the ridge and eaves, so is particularly suitable if a building is intended to face in a particular direction, e.g. towards the garden (for dwellings) or towards the street (for prominent public buildings).

Gable roof — Two juxtaposed inclined planes form a gable roof. This and the monopitch roof are the simplest pitched roof forms.

Mansard roof — A mansard roof has two juxtaposed roof planes of different pitches, and is now less commonly used. It was intended to give more headroom if the roof space were to be used.

If the end wall under the pitched roof areas is upright, it forms a gable. If this area faces a street or square, the building can be said to stand gable-on to the street. The opposite, eaves-on, is less common.

Hipped roof — If the roof slopes on all four sides it is known as a hipped roof.

Pavilion roof — A pavilion roof has all its roof planes pitched, with outside walls of equal length. The roof planes meet at a single point.

Half-hipped roof — If a roof has a gable and a pitched roof plane on the end wall, it is known as a half-hip.

Barrel roof — Roofs can be built with cylindrical vaulting, as a barrel roof. Roofs that are curved on all sides are domes.

Shed roof — Shed roofs have small monopitch roofs or gable roofs aligned like the teeth of a saw; the steeper plane is usually glazed. Fully glazed versions are common. They are often used to light large spaces such as production halls.

Fig.4:
Pavilion roof — Half-hipped roof — Barrel roof — Shed roof

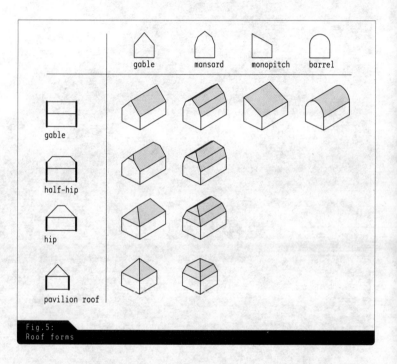

Fig.5:
Roof forms

\\Hint:
The term headroom defines the height of a space that people can use comfortably. An area is considered fully usable if it has a clear headroom of 2.2 m.

\\Hint:
When calculating the useful area of attic storeys, only half of any area 1 to 2 m high is counted. Areas 2 m and higher are counted in full, and areas under 1 m high are omitted altogether. Area calculations stipulations are derived from the individual national standards and areas where they apply.

The terms explained here are sometimes combined to describe a roof form fully, for example a hipped mansard roof. If the building has only one gable, this is not defined separately. The simple terms gable, mansard or monopitch roof are retained.

DORMERS

Dormers are also basic roof forms. They are used to light the roof space, or as an exterior design feature, providing additional usable space in the attic area. Note that far less incident light is admitted by the diagonal configuration of roof and dormer than by a window in a vertical wall. It is preferable to place windows in the gable walls. Dormer width should be restricted to one to two rafter fields. ＞ see chapter Roof structures Additional loadbearing elements may also be needed for the window; they can be tied into the frame. Note that the dormer pitch should not be less than the minimum roof pitch prescribed for the roof covering. ＞ see chapter Roof coverings All dormer surfaces must satisfy the same requirements of density, moisture, thermal insulation etc. as the roof itself.

Dustpan dormer

The form of the dormer depends on the roof covering. A dustpan dormer roof slopes less steeply than the main roof. Triangular vertical surfaces are created at the sides.

Gable dormer

A gable dormer also has vertical triangular areas, but meets the main roof at valleys, producing a street-facing gable roof.

Triangular dormer

The term triangular dormer is used only when there is a triangular gable planc.

Eyebrow dormer

The eyebrow rises out of the roof in a shallow arc. The roof plane is not interrupted. An eyebrow dormer is roughly ten times as wide as it is high. It requires the same roof height as a dustpan dormer.

Fig.6:
Dustpan dormer – Gable dormer – Triangular dormer

Fig.7:
Eyebrow dormer – Bull's eye – Barrel dormer

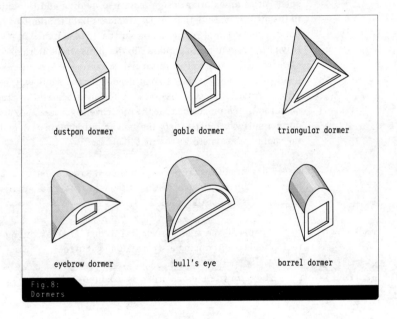

Fig.8:
Dormers

Bull's eye   The bull's eye forms a semicircle in the gable area. It can be fitted into various kinds of roof covering, but the circular shape requires a covering in sheet metal or some other flexible material.

Barrel dormer   If the semicircle is placed on a straight vertical area, the term barrel dormer is used.

Fig.9:
Wooden loadbearing structures

ROOF STRUCTURES

As well as roof forms, we distinguish between different roof constructions. For smaller roofs, intended for private houses, for example, wood is still the pre-eminent material. It absorbs compression and tensile forces well, and is reasonably priced and easy to work with on site. Loadbearing structures in steel or prestressed concrete beams are used when larger spans are involved; they can also be adopted for domestic building to achieve a particular design effect.

There are three basic loadbearing systems for pitched roofs: couple roof, collar roof and purlin roof.

Couple roof

The couple roof is a simple triple frame form: if the structure is viewed in cross section it consists of two beams leaning against each other, the <u>rafters</u>, connected to the floor below or to a <u>tie beam</u> to form a triangle. This triangular framework is called a pair of rafters. The beams are securely fixed at the connection points, but can turn freely, which is why such a frame is said to be hinged. A couple roof consists of several pairs of rafters in a row. They should be 70 to 80 cm apart, up to a maximum of 90 cm. The rafters are subject to loads from self-weight, snow etc. › see chapter Loads The tie beam linking them absorbs the tensile forces that are trying to pull the pair of rafters apart. Hence the connection between the rafters and the tie beam or ceiling must allow the forces generated to be transferred into the wall or supporting member below. In the traditional craft design, the tie beam projects beyond the triangular frame at the eaves. This projection is called the <u>verge member</u>.

Verge member

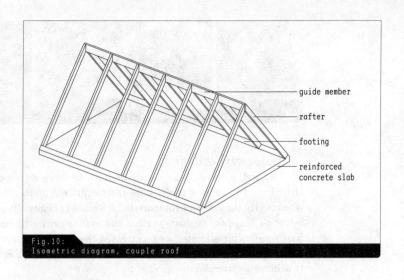

Fig.10:
Isometric diagram, couple roof

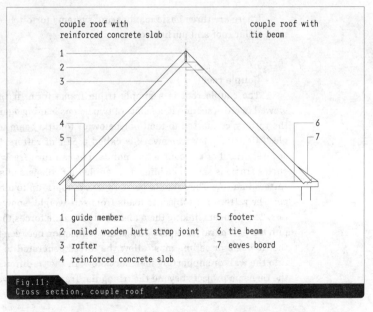

1 guide member
2 nailed wooden butt strap joint
3 rafter
4 reinforced concrete slab
5 footer
6 tie beam
7 eaves board

Fig.11:
Cross section, couple roof

Eaves board

The verge member will generally be more than 20 cm long. The roof pitch has to be extended here by adding an <u>eaves board</u> so that the roof covering can run out through the wall at the end of the building. This produces a curb or kink in the roof surface, and alters the pitch.

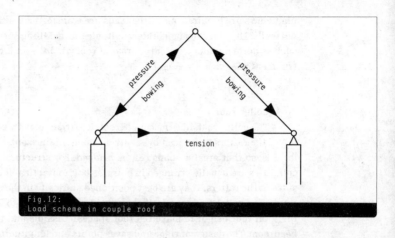

Fig.12:
Load scheme in couple roof

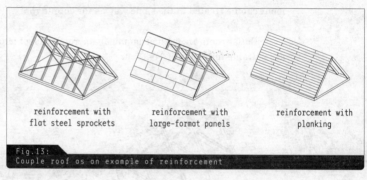

Fig.13:
Couple roof as an example of reinforcement

Nowadays, a projecting roof more commonly uses a ring beam, generally in concrete. Here, a spandrel beam or a floor bevel transfers the diagonal forces from the rafters into the ring beam.

Guide member

A <u>guide member</u> is commonly placed at the ridge joint to facilitate fitting the rafters. This links the rafters longitudinally, but the roof has to be sufficiently reinforced to take longitudinal forces. This can be achieved by fitting wooden sprockets or flat metal strips attached diagonally to the rafters, or by adding flat boarding.

Couple roofs may have a pitch of approx. 25 to 50°. They are suitable for spans of up to 8 m, as otherwise the requisite timber cross sections would not be economically viable. They tend to be used when column-free plan views are required, as all the loads are transferred via the longitudinal walls. They are only moderately suitable for installing special elements such as dormer windows or large areas of roof windows and the associated trimmers. › see chapter Trimming

Collar roof

The collar roof, like the couple roof, is a triangular framework. Bowing of the rafters is reduced by an arrangement of horizontal ties, the collar ties, so that greater spans can be bridged. For structural reasons, the collar ties are usually arranged in pairs, as horizontal ties, and fixed to the sides of the rafters. They are best positioned statically in the middle of the rafter. The collar ties can also be arranged at a height 65 to 75 percent of the total roof height to make the roof space accessible and provide more headroom. The design of ridge and eaves points and longitudinal reinforcement can be treated in the same way as for couple roofs.

Collar roofs are most economical at a roof pitch of more than 45°, and are suitable for spans of 10–15 m.

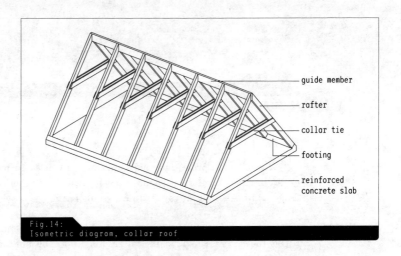

Fig.14:
Isometric diagram, collar roof

\\Hint:
A span defines the length bridged by an unsupported structural element. For couple and collar roofs this is the width of the building.

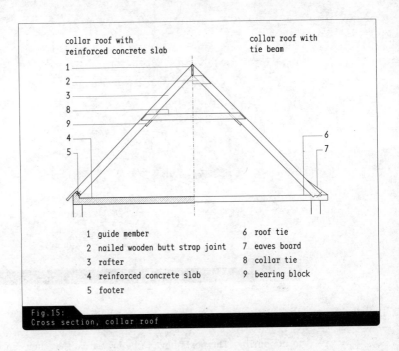

Fig.15:
Cross section, collar roof

1 guide member
2 nailed wooden butt strap joint
3 rafter
4 reinforced concrete slab
5 footer
6 roof tie
7 eaves board
8 collar tie
9 bearing block

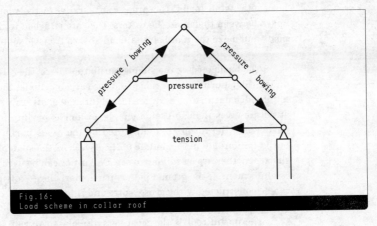

Fig.16:
Load scheme in collar roof

Purlin roof

Purlin roof

Post

A purlin roof has horizontal members – purlins – supporting the rafters. The purlins can be supported by the outside walls or by uprights, the posts, or stays. Rafters are subject to bending loads, which are transferred

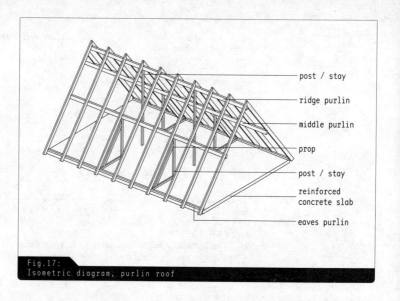

Fig.17:
Isometric diagram, purlin roof

Props

to the purlins. The posts have to be braced so that they can also absorb horizontal wind loads. The <u>props</u> run parallel to the roof pitch, with a space between them and the rafters. They are attached to the sides of the posts and make the roof better able to absorb transverse forces.

The basic form is the simple purlin roof. Here, the rafters are placed on the ridge purlin (at the ridge) and the eaves purlin (at the eaves). Loads in the ridge purlin area are transferred through posts. This simple support structure gives the expression <u>simple</u>, or single, <u>purlin roof</u>. In a double purlin roof, the rafters are supported by the eaves purlin and a central purlin (preferably in the middle of the eaves). As the span of the rafters is shortened, they are less liking to sag. Collar ties can be used for transverse reinforcement. If the ground plan is particularly large, a triple purlin roof can be constructed with eaves, centre and ridge purlins.

The purlin roof is the most versatile classical roof structure form. The rafter system is independent, and a wide variety of irregular and composite roofs can be constructed. Chimneys and windows can easily be fitted by trimming. › see chapter Trimming

Roof pitch can be selected at will for purlin roofs. Good rafter lengths are up to 4.5 m between the purlins.

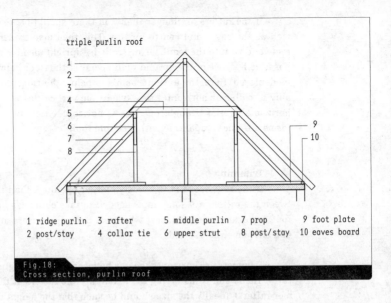

Fig.18:
Cross section, purlin roof

1 ridge purlin 3 rafter   5 middle purlin 7 prop   9 foot plate
2 post/stay   4 collar tie   6 upper strut   8 post/stay 10 eaves board

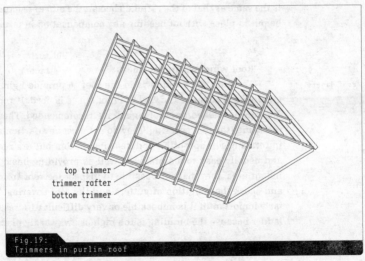

Fig.19:
Trimmers in purlin roof

It is important when using this structure that the loads from the posts can be transferred into the building's loadbearing system. The posts should be positioned on loadbearing walls, joists or upright members below them.

Truss

If the rafters are longer than 7 m there should be a ridge purlin as well as the eaves and centre purlins. This then also transfers forces directed down into the loadbearing walls by upright members. If this is not possible because of the ground plan layout of the floor below, a truss can be deployed. Here, the upright members below the ridge purlin are taken only to collar or horizontal ties, where the vertical loads are transferred horizontally. The triangle of rafters, horizontal ties and uprights then forms a frame comparable with a couple roof.

Trimming

Trimmers

If the roof space is to be used, larger openings than the spaces between the rafters are sometimes needed to take chimneys up through the roof, for example, or to insert windows. Here, the rafters are interrupted and a trimmer is inserted.

Trimmers are horizontal beams between rafters bridging one or more rafter fields and transferring the loads into the rafters at the sides (trimmer rafters). Ideally the size should be such that the height of the timbers is the same as that of the rafters. Then all the other structural elements can be put in place without needing any compensation in terms of level.

Roof windows and dormers

Roof-lights

The placing of windows in the roof to provide light and ventilation must take account of the way the roof space is used. If roofs are not used but can be walked on, roof-lights are recommended. These are fitted directly into the roof covering. They do not meet any particular requirements in terms of windproofing and heat insulation, but are reasonably priced and usually easy to maintain. As well as providing basic lighting for the indoor area, they make it possible to inspect the roof from the roof space and check the condition of gutters and the roof covering. This is particularly important if it is impossible or very difficult to access the roof with a ladder because the building is too high or awkwardly placed.

Adequate lighting is essential if living or working accommodation has been built into the roof. As a rule, the proportion of window area should be at least one eighth of the ground area. Façade windows in the gable (if there is one), dormer windows, and windows in the roof itself can be used to provide light and ventilation. The advantage of vertical windows in dormers and gables is that they do not easily get dirty, as precipitation does not usually fall directly onto the surface of the glass. At the same time, it is possible to come right up to the windows and so enjoy a better

> \\ Hint:
> The minimum specifications for roof windows as escape routes are laid down in the legislation. In most cases a window 90 cm x 120 cm is considered large enough. The parapet height should be a maximum of 120 cm, so that it is possible to climb over it from inside. The windows should not be further than 120 cm from the eaves. Otherwise a rescue platform must be fitted, so that the people concerned can attract attention.

view of the surroundings. Concerning possible fire escapes routes, care should be taken that roof windows can also be used as a "second escape route" (the first is down a staircase). It case of fire it is essential that the roof windows are large enough, and that people in the roof space can attract attention to themselves.

For roof space ventilation, care should be taken at the planning stage to allow for the fact that that heat will rise. If the roof space is open to the ridge, measures should be taken to ensure that hot air can escape from the upper roof area.

Windows in the roof

Windows built into the surface of the roof itself are now usually installed by the roofing contractor as prefabricated elements. All the connecting parts in the various makes are available for different coverings and rooftop fittings. The products available conform to the usual distances between rafters (between 70 and 90 cm). Roof windows can be installed between two rafters, or in one rafter field, or can extend over several rafter fields, in which case a trimmer must be fitted. > see chapter Trimming If rows of windows are needed, combination covering frames are available that enable several windows to be arranged next to each other without trimmers. The height of the windows is restricted to approx. 1.6 m, as otherwise it is impracticable to open or close them, and they weigh too much. If a larger area has to be fitted with windows, two can be installed, one above the other. The top edge of the windows should be at least 1.9 m, so that it is possible to look out. The parapet height is usually between 0.85 and 1 m. A higher parapet could be used for kitchens and bathrooms. The lower the roof pitch, the greater the area of window needed to achieve the necessary height. When planning roof windows, care should be taken not to break the ventilation cross section inside the roof structure. The air must be able to circulate freely around the window. Sheeting strips must be attached to the windows on all sides, and any water must be able to flow away unimpeded (e.g. by installing a wedge above the window). The heat insulation must be taken right up to the roof window and the gaps must be completely filled with insulation. The vapour barrier must also be attached on all sides, so that atmospheric moisture in the room cannot penetrate the insulation.

We distinguish between various opening methods for roof windows. A <u>swing window</u> is fixed in the middle at the sides. When opening it, the upper half swings into the roof space, while the lower rises outside it. A <u>hinged window</u> is fixed at the top edge and the whole window folds upwards and outwards. As it is very hard to clean the outside of such windows (by reaching around the window), combined <u>folding</u> and <u>swing windows</u> are often used. Hinged windows are also offered with sliding functions. Here, the window is pushed away to the side so that it then lies above the surface of the roof. Such windows are very elaborate and expensive, and so seldom used.

Care must be taken when fitting roof windows to leave sufficient distance between the windows and the adjacent buildings or boundaries. Building regulations lay down that roof windows must be approx. 1.25 m from firewalls. For terraced houses set gable-on to the street, › see chapter Roof types the windows must be at least 2 m from the eaves. Roof windows are often used when building regulations do not permit dormers.

Dormers

Lighting the roof space with dormers creates additional usable space, because they allow more headroom. The interior height under the dormer must be at least 2 m. Just as for roof windows, the parapet height is between 0.85 and 1 m. Small dormers can be placed above a rafter field, and are then the same width (approx. 70 to 80 cm). Trimmers have to be used for wider dormers. For couple roofs, this is usually possible only if the dormer width is restricted to two rafter fields. The rafters in purlin roofs can be interrupted in several fields. The forces from the rafters are then transferred into the ceiling vertically.

The front of the dormer is made from a squared timber frame, known as a <u>dormer truss</u>. It is placed either on the rafters or directly on the intermediate floor. If the dormer truss is placed on the intermediate floor, the roof space between the dormer truss and the eaves is usually sealed off and concludes the space. For shed dormers the key rafter is placed directly above the trimmer rafter. It is then supported by the dormer truss at the dormer end. The side triangle is called the <u>dormer cheek</u>. This area is usually reinforced with rough tongue-and-groove sheets or some other flat material. The side surfaces of the dormers must be insulated and sealed just like the rest of the roof. The dormer cheeks are usually clad in metal or slate because these materials are good for application to inclined surfaces. It is also possible to glaze the sides of dormers, which provides correspondingly more light.

The width of the dormer is determined by the coverage span of the material concerned. This is calculated from the width x number of pantiles (or other covering material). The equivalent applies to the length of the area removed. ᐳ see chapter Roof battens

If time is short, fitting prefabricated dormers is recommended. They can usually be placed on the existing roof within a single day, and the roof can be watertight again within the shortest possible period.

LAYERS OF STRUCTURAL ELEMENTS

Roof coverings

The roof covering's principal function is to allow precipitation to drain away reliably and to prevent moisture penetration from driven snow, for example. Roof covering should be rain- and weatherproof, and also fireproof. They must also guarantee moisture transfer from the inside to the outside, and protect the structural elements underneath them from the wind. Key features in the choice of a suitable covering are design, and then the roof pitch and the shape of the roof. Valleys or angles are more easily created with small-format materials such as flat tails. ᐳ see section Flat elements Large straight areas of roof are more easily and economically created with pantiles. However, many coverings work only up to a certain minimum roof pitch. The manufacturer's stipulated roof pitch for a particular roofing material always relates to the minimum pitch unless otherwise stated. If this is not reached in some places, water or dust can be prevented from penetrating the structural or insulation course by an underlay. ᐳ see section Waterproofing Various types of roof covering material are available, again in different materials.

Thatched roof

One ancient form of covering that is now found only regionally or sporadically is reed or straw thatch. It should be applied at an angle greater than 45°. At an ideal 50° the wind presses the thatch against the substructure and the proportion of lifting force is low. The covering is attached to a framework in several superimposed bundles.

Flat elements   Flat roof covering elements can be in wood (shingles), stone, concrete or clay. The standard roof pitch for shingles depends on the length of the shingles, the overlap for the individual shingles and the number of courses. Simple two-course shingle roofs can be constructed only if the pitch is a minimum of 70°, in other words almost horizontal, while the more elaborate three-course version goes to 22°.

Flat stone elements are usually made of slate. They come in the form of rectangular, acute-angled, scalloped or scale tiles and are pinned to a framework in the overlay area, following individual rules. The standard roof pitch for slate coverings is 25 to 30°.

Concrete or clay tiles are produced industrially. The advantage is that they can be bought to suit specific situations, such as edging or penetrations. An upstand can also be created if the tiles are to be laid on roof lathing. The standard pitch for concrete and clay tiles is between 25 and 40°.

Profiled tiles   Profiled roof tiles in their various forms are made of clay or concrete. Just like the industrially produced flat tiles, special shapes can be prefabricated for many particular situations. Unlike flat tiles, profiled tiles overlap on three sides.

Under-and-over tiles   The oldest profile tile is the under-and-over tile: conical hollow tiles are placed so that they interlock. The upper tile is concave, and takes the water into the lower, convex, tiles, which drain the water into the gutter. They have no rims or ribs. Modern tiles' upstands are shaped to interlock at the top and sides, to prevent water penetration. The standard roof pitch for profiled tiles is 22 to 40° according to type.

There are also profiled tiles, such as corrugated tiles, available in various materials for large-format roof coverings. The fibre-cement

> \\Hint:
> Precise details about standard roof pitches for certain roof coverings can be found in more advanced literature, for example in "Roof Construction Manual – Pitched Roofs" by Eberhard Schunk et al., published by Birkhäuser Publishers, or in the manufacturers' details.

Fig.20:
Thatched roofs

Fig.21:
Flat roof coverings: wooden shingles

Fig.22:
Flat roof coverings: slates

Fig.23:
Flat roof coverings: flat tiles

Fig.24:
Roof coverings using profiled tiles

corrugated tile is a simple example. These are laid overlapping on battens, in runs about 1 m wide. Various manufacturers supply versions for edges, intersections and upstands. They can be used for roof pitches of less than 12°. Lower roof pitches are permissible for corrugated bitumen roof coverings. Here, edges or connections are usually constructed using sheet-metal angles.

Industrial construction

Some profiled metal elements can even be used for standard roof pitches of up to 5°. These metal coverings are usually made of galvanized steel, copper or aluminium alloys. They are laid as corrugated profile sheets, following the same principles as fibre cement or bitumen coverings, or as trapezoidal profile sheets. Trapezoidal metal sheets are available in various shapes and sizes. They are made of thin, folded metal sheeting. The edges are optimized in terms of loadbearing properties and can carry loads over long spans. Trapezoidal sheets are supplied as composite sheets with thermal insulation, for industrial construction in particular. The edges of the sheets must overlap and interlock through upstands in order to guarantee the roof's impermeability. They are fixed to the supporting battens with screws, bolts or clips. When working with metal coverings it is important to ensure that no contact corrosion with other metals ensues. A separating course should therefore be placed under the covering ⟩ see chapter Flat roofs, Layers of structural elements if the purlins are made of a different metal, or of concrete.

Strips

Strips are another form of metal covering. They are made of lead, aluminium, copper or stainless or galvanized steel. The strips are usually 500 to 1500 mm wide. They are laid in rows, or courses. The side edges are joined with a welt, a roll or an overlap. The horizontal ends of the sheets are finished with overlaps or transverse welts. Connections to other structural elements or ends are created by hand from turned-over sheet metal. The standard pitch for this kind of roof covering is 5°. Additional precautions are nevertheless recommended for lower pitches.

Roof battens

Flat roof covering materials are fixed to the battens with screws, nails, bolts or clips, but roof tiles with an upstand are laid on battens.

Roof battens

The dimensions of the battens depend on the weight of the covering and the rafter spacing. The following longitudinal cross sections are recommended for average covering: up to 30 cm between rafters -> 24/48 mm battens; 80 cm between rafters -> 30/50 mm battens; 100 cm between

Fig.25:
Roof coverings using under-and-over tiles

Fig.26:
Roof coverings using profiled tiles

rafters -> 40/60 mm battens. The quality grading of the timber should also be taken into account.

Calculating spacing between rafters

The horizontal spacing between the rafters depends on the roof pitch and the choice of roof covering elements. Different values apply according to manufacturer and product. The length of the area to be covered must first be established. It is roughly equivalent to the length of the rafters. The bottom edge of the roof is a special case. The planner must decide, in relation to the structure as a whole, whether the last row of roof tiles should project beyond the ends of the rafters, finish flush or even conclude with an upward tilt. The usual choice is an overhang, corresponding with the overlapping tile area. This is defined as the <u>undereaves course dimension</u>. At the ridge, the covering overlap on the roof batten requires a space to be kept free. This is defined as the <u>ridge course dimension</u>. To calculate the number of rows needed and the resulting space between the battens, the ridge and undereaves course dimensions must be subtracted from the overall length to be covered. The remaining length is divided into equal spacings with a tile overlap as prescribed by the manufacturer.

The number of tile elements needed for the breadth of the roof is also calculated on the basis of the manufacturer's approved dimensions. Here, too, the spacing can very slightly to allow for tolerances and insertions such as chimneys or pipes.

Fig.27:
Roof battens

Table 2:
Roof batten cross sections

Batten cross sections in mm	Axis width in m	DIN 4074 timber grades
24/48	up to 0.70	S 13
24/60	up to 0.80	S 13
30/50	up to 0.80	S 10
40/60	up to 1.00	S 10

A statical examination should be conducted for all other spacings between rafters or batten cross sections.

Cross battens

Cross battens are needed where there are flat tile underlays, or for roof pitches of less than 22°. The same applies to intermediate roof coverings or support systems › see chapter Waterproofing where moisture cannot drain off freely.

\\Example:
Specimen roof batten spacing calculation:
Overall length to be covered: 7.08 m
Undereaves course: 32 cm (to be established)
Ridge course: 4 cm (according to manufacturer's instructions)
Length to be divided: 7.08 m – 0.32 m – 0.04 m = 6.72 m
Average batten spacing according to manufacturer's instructions: 0.33 m
6.72 m : 0.33 m = 20.4 rows
Number of rows selected: 20 rows
It is necessary to check whether the chosen batten spacing of 0.33 m lies within the manufacturer's permitted parameters for the particular roof pitch.

\\Tip:
Cross battens are not absolutely essential with underlays. In such cases the underlay is fitted with a slight sag, so that any water that may penetrate can drain off safely under the roof battens. Even here, cross-battening is recommended, as most underlays shrink over time, and so the material stretches.

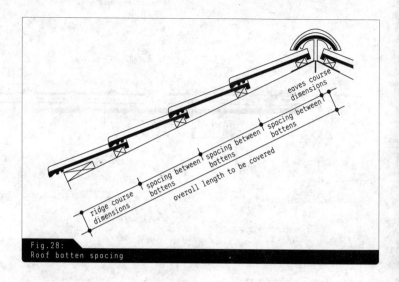

Fig.28:
Roof batten spacing

Waterproofing

The roof covering usually ensures that the roof is rainproof. However, in particularly acute situations, additional planning measures should be taken to obviate spray penetration in cases of high wind or driven snow. Acute situations can be caused by an unduly low roof pitch, highly structured roof surfaces, special roof forms, the adaptation of the roof for living space etc. Special climatic conditions can also make additional measures necessary; for example an exposed position, or areas subject to high winds or frequent heavy snow.

Underlay

The simplest additional element is an underlay. This is fitted as a ventilated sheet structure, i.e. the sheeting is not supported below, but hangs freely between the rafters. Underlay is supplied in rolls, and is usually in the form of reinforced plastic sheeting.

Supported layer

Here the sheets are laid over a support, such as a timber structure. Different qualities are achieved through the nature of the seams. Such supported layers are classed as rainproof. They are fitted below the battens and counter battens.

Sewn welded underlays

These consist of waterproofing sheets joined by welding or gluing to make them waterproof. We distinguish between rainproof and waterproof underlay. A rainproof underlay may include structurally required apertures. The sheets are positioned under the battens and cross battens. No

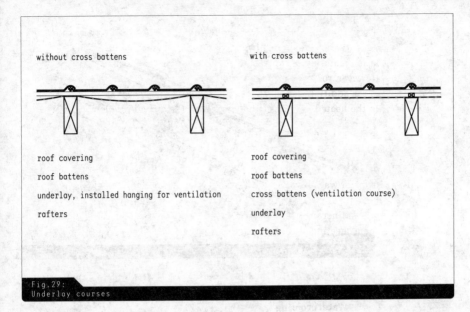

Fig. 29:
Underlay courses

Table 3:
Extra requirements for roof coverings

Roof pitch	Special requirements relating to use, structure or climate			
	No special requirements	One additional requirement	Two additional requirements	Three additional requirements
> standard roof pitch	-	Underlay	Underlay	Supported overlapping course
> 6° below standard roof pitch	Underlay	Underlay	Supported overlapping course	Separate welded / glued course
> 10° below standard roof pitch	Rainproof underlay	Rainproof underlay	Rainproof underlay	Waterproof underlay
< 10° below standard roof pitch	Rainproof underlay	Waterproof underlay	Waterproof underlay	Waterproof underlay

apertures are permitted in a waterproof underlay. The counter battens are an integral part, i.e. the sheets are fitted between the battens and counter battens. Close attention must be paid when fitting to ensure that water accumulating on the sheets can drain away into the gutter at the eaves. Adequate ventilation must be provided in the roof space under the underlay.

Insulation

If roof spaces are ventilated and not used as living space, the building below is usually insulated on or under the last ceiling. This saves insulation material, in contrast with an insulated roof, and is easier to install. However, roof spaces are increasingly being developed to take advantage of the additional space they can provide. In such cases the roof is included in the heated area of the house and all the structural elements enclosing the used section must meet the appropriate thermal insulation standards.

> see Appendix

It is important to ensure that the insulation course joins with the exterior wall insulation course, to avoid cold bridges. > see Fig. 31

Thermal insulation

Mineral wool, rigid polystyrene foam boards (PS), rigid polyurethane foam sheeting (PUR), cork, lightweight wood wool sheets and pouring-type granular insulation material are available for thermal insulation. The individual insulation materials are offered with different thermal conductivity ratings. Insulating materials with particularly poor thermal conductivity insulate correspondingly well, and can be thinner than materials with good conductivity, while still providing the same level of insulation. Thermal conductivity, the transmission constant (k), is given in watts per square metre Kelvin (W/m^{2}K). The lower the value, the better the insulating properties.

Thermal insulation can be fitted between the rafters, which must be high enough to provide a sufficiently thick insulation course. A higher rafter than statically necessary to provide the necessary space may be selected, or an additional insulating course can be fitted to the rafters from inside the space. The timber used for the rafters is such a poor conductor

---

\\Hint:
Thermal bridges are particular points in the structure where the building's insulation course is interrupted. Here heat can penetrate the building when there are temperature differences between inside and outside (summer conditions), or it can escape (winter conditions). Situations of this kind should be avoided where possible as water may condense on cold surfaces when the air cools, be trapped inside the structure and damage it.

\\Tip:
Most insulating materials are available in widths to match the most common spacing between rafters, so information about products should be gathered at an early planning stage. It is recommended that the distance between the rafters be fixed according to the gap required, rather than by an even unit spacing.

Fig.30:
Thermal insulation of pitched roofs

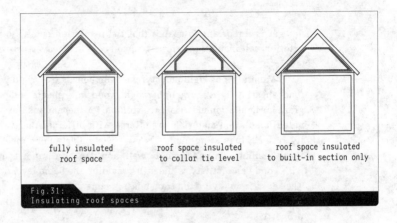

fully insulated roof space

roof space insulated to collar tie level

roof space insulated to built-in section only

Fig.31:
Insulating roof spaces

that thermal bridges are not created even if the insulating course is interrupted.

Insulation between rafters

Full rafter insulation
Insulation on rafters

Vapour barrier

Insulation between rafters means that there is an air space between the thermal insulation course and the underlay sheeting, to ventilate the structure. If the height of the rafters is exactly the same as the thickness of the insulating material, the term full rafter insulation is used. If the rafters are to remain visible on the inside, the insulation can be fitted to the roof on the outside, with boarding, as insulation on rafters.

A vapour barrier is essential for thermally insulated roofs. It is always positioned on the inside of the space, below the thermal insulation. It must cover the full area, and be attached at the edges in such a way that it is airtight. Vapour barriers prevent moisture in the air inside the space from penetrating the thermal insulation or the roof structure through diffusion.

The appropriate insulation rating for the vapour barrier sheeting is determined according to the roof pitch and the length of the rafters. The

> **\\Hint:**
> Any air from inside the space that might diffuse through the thermal insulation would cool down as it passes from the inside to the outside (in winter). As warm air contains more moisture than cold air, water could condense, making the thermal insulation and the structure damp, and thus damaging the building.
> If several vapour barriers are fitted inside part of a building, the outermost in each case must be more permeable to water vapour than the inner barriers. Any moisture that may accumulate can thus escape to the outside.

> **\\Tip:**
> Solid building materials can be used instead of vapour barriers, provided they render the insulation airtight in relation to the interior space. These can be OSB boards (Oriented Strand Boards), for example, with their joints sealed so that they are airtight (using appropriate adhesive tapes). Evidence of effectiveness is usually required here.

vapour insulation rating gives the equivalent diffused air space depth (s_d) for the layer of air. This measures the resistance the material offers to water vapour transmission. Vapour barriers can be fitted in elastomer sheeting or in plastic versions.

To sum up: the structure, and the thermal insulation course in particular, must be protected against water penetration. This is achieved on the inside by fitting vapour barriers, which prevent penetration by moisture in the air. Outside, moisture in the air can evaporate above the ridge underlay or through the roof covering. Any water that may accumulate is drained into the gutters via the underlay sheeting.

TYPES OF FINISH

Unventilated roof structure

Unventilated roof structures are constructed with full rafter insulation. The insulation is fitted between the vapour barrier (on the inside) and the underlay (on the outside). Counter battens should be placed on top of the underlay, to ensure that any moisture that may penetrate into the gap can evaporate under the roof covering. This construction method is used when the structural elements must be kept as thin as possible, or when a roof is having a structure inserted subsequently, and there is insufficient rafter height for a ventilated structure. But if ventilation for the thermal insulation is considered essential, the rafters can be doubled by fitting an additional lath to the rafters from the outside or by installing a second thermal insulation course inside.

Ventilated roof structure

If this approach is used, air is able to circulate below the underlay (roof membrane). The advantage is that any moisture present can evaporate from the insulation material and escape via the ridge joint. The air space also means that the less heat is absorbed into the roof space (in summer). The warm air rises and escapes through the ventilation apertures in the ridge. A suction effect is produced, as in a chimney. It is essential here that sufficient air can flow in through apertures in the eaves.

When planning the air space depth, care should be taken not to fit the insulation completely flat, so that it has room to swell subsequently. If the air is to circulate freely, it must be diverted around trimmers, roof windows, chimneys or dormers, so counter battens should be fitted, enabling air to circulate between the eaves and the ridge. ˃ see Fig. 29

Supported thermal insulation

Structures with supported thermal insulation are not usually ventilated. Instead, the roof is covered with prefabricated elements supported by falsework above the rafters. Prefabricated elements of this kind are usually supplied with a vapour barrier on the underside. As in full rafter insulation, the underlay or an appropriate material for draining water away is fitted directly on top of the thermal insulation. This finish reduces building time. The rafters can remain visible underneath, and be used as a design element. In this case the rafters should be finished with planed facing timbers, or with laminated timbers.

Internal surfaces

The internal surfaces of developed roof spaces should be clad with a material that can absorb moisture and release it again, to produce a pleasant atmosphere in the space. Wall rendering usually performs this function in spaces with masonry walls. Plasterboard sheets are customarily

\\ Hint:
The ventilation cross sections for a ventilated roof structure should correspond to the following minimal values: for roof pitches of more than 10°, 2‰ of the roof area at the eaves, but at least 200 cm², and 5‰ of the roof area at the ridge. The free ventilation cross section must be at least 2 cm (air space depth). For roof pitches of less than 10° the ventilation cross section must be 2‰ of the total roof area at two opposite eaves. The free ventilation cross section must be at least 5 cm. Precise specifications can be found in national standards.

insulation between rafters

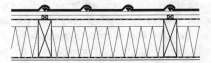

roof covering
battens
counter battens
underlay
air space
thermal insulation
vapour barrier

insulation between rafters with additional internal insulation

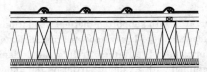

roof covering
battens
counter battens
underlay
air space
thermal insulation
vapour barrier

full rafter insulation

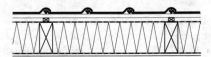

roof covering
battens
counter battens
underlay
thermal insulation
vapour barrier

insulation on rafters

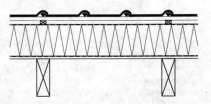

roof covering
battens
counter battens
underlay
air space
thermal insulation
vapour barrier
falsework

Fig.32:
Types of structures for pitched roofs

Fig.33: Drainage

used for building inside the roof space. They are easy to work and adapt on the building site. Note here that slight movements in the structure can easily produce cracks. Another common wall cladding is produced using matchboard with tongue and groove. This will not crack easily, because the boards can shift in relation to each other. But in most cases the principal deciding factor when choosing interior wall cladding is appearance.

DRAINAGE

Pitched roofs are drained towards the eaves via the roof surfaces and valleys. The precipitation water collects in gutters there, and is taken into the sewerage system via downpipes. The term <u>combined system</u> is used when the water feeds into the public sewerage grid, and <u>separate system</u> when the precipitation feeds into the local groundwater.

Dimensioning

The assumed local rainfall load must be ascertained in order to dimension the gutters and pipes.

The rainfall is calculated from the local rainfall load – generally at 300 litres per second per hectare ($r = 300 l/(s*ha)$) – the runoff coefficient, which takes the roof pitch and the nature of the surface into consideration, and the base roof area.

The calculation uses the formula

rainwater runoff (in l/s) = runoff coefficient × precipitation area serves (in m^2) × design rainfall load.

The value produced can then be used to select the appropriate downpipe from standards tables.

Gutters

Gutters are fitted to the eaves with height-adjustable gutter brackets. One bracket per rafter is fitted for timber roof structures, but the brackets

Table 4:
Runoff coefficients (extract from DIN 1986-2 Table 16 / ISO 1438)

Surface types	Runoff coefficient
Surfaces impermeable to water	1.0
Roofs with a pitch of <3°	0.8
Gravel roofs	0.5
Green roofs	0.3
Extensive green roof (d>10cm)	0.5

Table 5:
Precipitation areas that can be attached to rainwater downpipes of minimum pitch with precipitation quantities of 300 l/(s*ha) according to DIN 1986-2 Table 17 / ISO 1438

Roof pitch	Maximum admissible ar in l/s	Runoff coefficient 1.0 r in m^2	Runoff coefficient 0.8 r in m^2	Runoff coefficient 0.5 r in m^2
50	0.7	24	30	48
60	1.2	40	49	79
70	1.8	60	75	120
80	2.6	86	107	171
100	4.7	156	195	312
120	7.6	253	317	507
125	8.5	283	353	565
150	13.8	459	574	918
200	29.6	986	1233	1972

Table 6:
Design for downpipes and gutter assignment (here for PVC) according to DIN 18 461 Table 2 / ISO 1438

Roof area to be attached at a maximum rainfall load of 300 l/(s*ha) in m^2	Rainwater runoff in l/s	Rainwater downpipes (nominal dimension) in mm	Gutter assigned (nominal value)
20	0.6	50	80
37	1.1	63	80
57	1.7	70	100
97	2.9	90	125
170	5.1	100	150
243	7.3	125	180
483	14.5	150	250

should never be further than 90 cm apart, depending on the structure. The gutters are laid in the gutter brackets. Care should be taken here that the gutter slopes outwards, i.e. that it is higher on the building side than on the outside. Any water that may overflow is thus directed away from the building. The gutter should slope (minimum 2%) towards the drainpipes. Metal gutters in particular expand or contract if temperatures vary, so lengths of 15 m should not be exceeded. Individual gutter sections are joined using connectors. Stop ends are fitted at the ends of the gutters to seal them off.

Downpipes

Gutters have prefabricated joints to which the downpipes are attached with elbows. Sections of pipe are connected by waterproofing and connecting sockets. The pipes are attached to the building by brackets with pins or screws. The pipe should clear the building by more than 20 mm so that damp will not penetrate the wall if the pipe is damaged.

Gutters and pipes can be either angled or round. Various materials are available. Care should be taken that the materials cannot interact with each other to produce corrosion, for example, or create tension as a result of different expansion properties. So, for example, copper gutters and pipes can be assembled only with copper-clad steel brackets and clips. Brackets in galvanized steel or aluminium are recommended for aluminium gutters. Galvanized steel brackets and clips are available for zinc or galvanized steel gutters. PVC gutters can be fitted with galvanized steel or plastic-clad brackets.

It is also recommended that foliage interception grids be fitted to the gutters. These are supplied in the form of longitudinal baskets curving outwards, and increase cleaning intervals. Heated guttering can be installed at awkward places, such as rising structural elements around an internal gutter › see Hint page 63; these will guarantee that water will drain off even in case of snow.

Internal guttering

Internal guttering is the term used for gutters that are not suspended from the eaves, but positioned above the floor slab. › see Fig. 33 left This design

\\Hint:
Local authorities can supply values for areas with heavy or light precipitation. An additional 100% safety element should allowed for internal gutters.

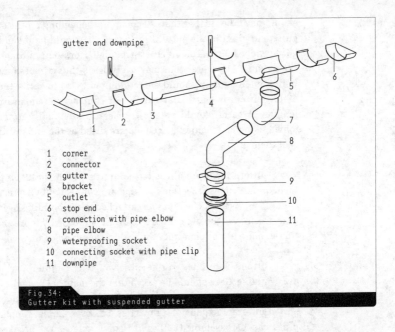

Fig. 34:
Gutter kit with suspended gutter

1 corner
2 connector
3 gutter
4 bracket
5 outlet
6 stop end
7 connection with pipe elbow
8 pipe elbow
9 waterproofing socket
10 connecting socket with pipe clip
11 downpipe

is chosen if a roof overhang is not desired or not admissible. A safety gutter can be included so that even if it leaks, water cannot penetrate the building. This can take the form of a waterproofing strip under the metal or plastic gutter. An emergency overflow spout can also be provided.

PRESENTATION

Buildings are presented as drawings in plan views, sections, elevations and details. When drawing a plan view, a notional section is taken at a certain level (usually at 1 to 1.5 m above the finished floor of a particular

> \\ Hint:
> Individual dimension units are particularly important for working plans. Building dimensions and structural element lengths are usually given in metres (in centimetres for smaller dimensions where appropriate). Dimension chains are compiled in metres or centimetres. Structural element cross sections are given in centimetres. Steel structural elements are dimensioned in millimetres.

Rafter plans

storey), so only the structural elements below this are shown. It is therefore often difficult to provide adequate details of the roof structure.

For the above reason, a rafter plan is usually prepared for pitched roofs. It shows all the wooden structural elements (or the corresponding features for other structures) as a top view. The roof covering, waterproofing, internal cladding etc. are not shown. The plan is used as a working guide by the on-site carpenters. In order to define the position of the roof structure, also called roof truss, in the building, the topmost floor slab must also be shown (unbroken line for visible structural element), along with the walls below (dashed line for concealed structural element), and a dimensional reference to the roof. If structural elements such as supports, stays or angle braces are concealed by the rafters above them, they are shown as dashed lines. The roof battens and counter battens are not shown, even though they are timber structures, because they will be later fitted by the roofers, not the carpenters. All the different wooden parts are allocated a position number, which is usually entered on the plan as a circled number, with a line or arrow to connect it to the particular structural element. A legend then describes the structural elements. Information must be given about the dimensions (width times height of cross section) and grade of the building material.

Information about installation or connections should also be recorded. The structural elements are dimensioned by the architect, structural designer or structural engineer.

In addition to the rafter plan, the following details should be provided for the building work, with drawings of all the layers of structural elements: ridge joint, eaves, verge, penetrations, apertures, and all special solutions.

\\ Hint:
The dimensions of the wooden structural elements are usually given as cross sections. A beam with a rectangular cross section of 10 cm x 12 cm is defined as 10/12, spoken as "ten to twelve". Round parts are defined by their diameter, and steel parts by their product or profile designation (e.g. HEA 120, U 100 etc.). Length and position are fixed according to the dimensions in the plan.

\\ Hint:
Further information about drawings can be found in "Basics Technical Drawing" by Bert Bielefeld and Isabella Skiba, published by Birkhäuser Publishers, Basel 2007

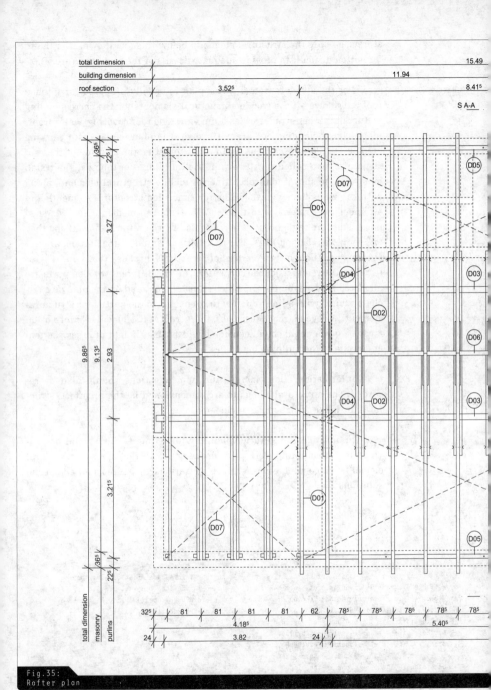

Fig.35:
Rafter plan

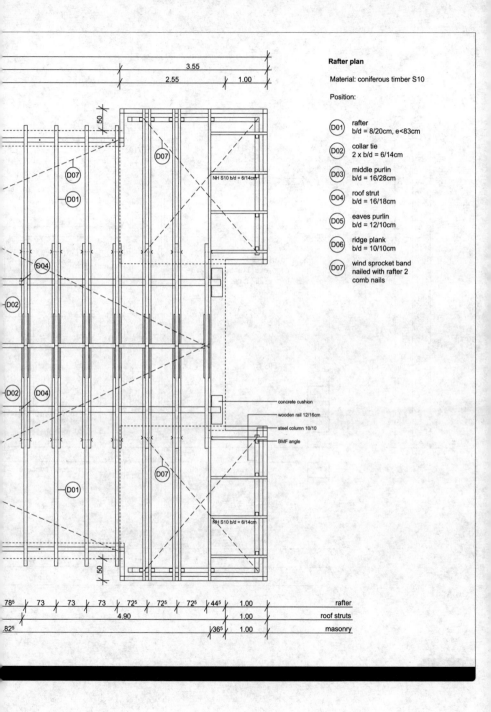

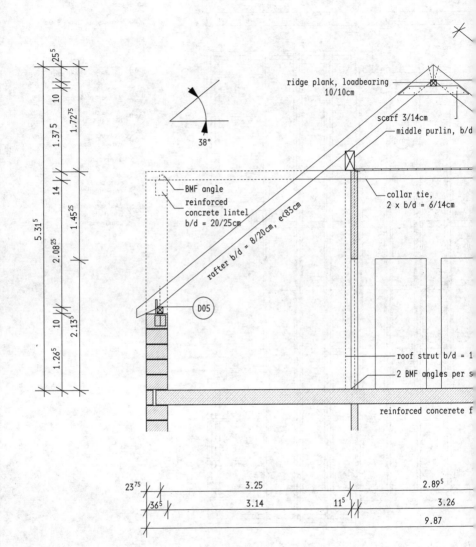

Fig.36:
Rafter plan, section

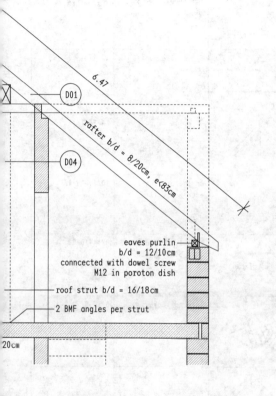

SECTION A-A

Material: coniferous timber S10

Positions:

- (D01) rafter
  b/d = 8/20cm, e<83cm
- (D02) collar tie
  2 x b/d = 6/14cm
- (D03) middle purlin
  b/d = 16/28cm
- (D04) roof strut
  b/d = 16/18cm
- (D05) eaves purlin
  b/d = 12/10cm
- (D06) ridge plank
  b/d = 10/10cm

Table 7:
Summary of structural elements for pitched roofs

Structural elements	Illustrations	Hints	Usual dimensions for average roofs
Eaves board		For couple roofs	8/12–10/22
Tie beam		For couple roofs	12/12–14/14
Ridge purlin		On walls or posts	14/16–16/22
Eaves purlin		On the ceiling or external wall	10/10–14/16
Collar tie		Usually in the form of a binding tie	8/14–10/20
Collar rafter		Internal curb	8/14–8/22
Angle clip		For tie bars	
Angle brace ties		Strut on post in longitudinal direction	10/10–10/12
Middle purlin		Under the rafters	12/20–14/20

Table 7:
Summary of structural elements for pitched roofs

Structural element	Illustration	Hints	Usual dimensions for average roofs
Post		Supports purlins	12/12–14/14
Guide member		For simple assembly	Thicknesses from 22 mm
Rafter		Supports roof covering	8/14–8/22
Brace		For transverse reinforcements	14/16
Verge member		For couple roof	Length 20 cm
Trimmer		For apertures	8/14–8/22
Sprocket		Reinforcement	In flat steel
Binding tie		Horizontal reinforcement, in pairs	6/14–8/16

FLAT ROOFS

BASICS

A flat roof has a pitch of less than 5°. However, flat roof structures are suitable for shallow-pitch roofs (up to 25°), although here the building materials must be adequately protected against slippage. The top section of a flat roof is the roof covering, known as the roof weatherproofing or waterproofing. Flat roofs can be constructed to support people or vehicles. For example, the space on the roof can be used as a terrace, parking deck or for fitting ventilation equipment. The roof surface must be carefully planned: seen from the neighbouring buildings the roof is the fifth façade, and is therefore important in terms of the building's appearance as well. Like pitched roofs, flat roofs have to meet various requirements. They must provide adequate protection against moisture and precipitation, as well as against heat, cold and wind. They must transfer the forces exerted by their self-weight, snow and traffic safely into the structure below them. Wind suction has an important part to play as a load, as the materials used to waterproof the roof are usually light, and thus have to be secured against lifting and mechanical movements.

Flat roofs can be finished flush with the walls, or they can protrude. The upturn at the edge of the roof is known as the roof parapet. It is where the insulating and waterproofing elements of the roof are brought together. Precipitation is usually drained into the interior of the building and thus directed from the edge of the roof towards the roof outlets. An incline of at least 2% must be guaranteed.

LAYERS OF STRUCTURAL ELEMENTS

Flat roofs can have various functional layers, which must be matched to each other. It is essential that the layer sequence provide adequate insulation against heat and noise.

> \\Hint:
> Evidence about heat and noise insulation for the whole building, according to the building type, is part of the planning permission process with the responsible building department.

The moisture to which the structural layers are exposed (e.g. trapped humidity, condensation build-up) must be able to escape. Building materials must be compatible with each other.

Undercourse

The course below the waterproofing, which deals with water, is called the undercourse. It can be the loadbearing structure, e.g. a concrete surface, or also boarding or the thermal insulation component. It is important that the waterproofing material match the undercourse, to avoid expansion cracks, for example. Joints between prefabricated parts must be covered by separator strips at least 20 cm wide.

Bonding course

Bonding courses are installed to improve the adhesive properties of building component layers. These can be primers or prior applications of bitumen solutions and bitumen emulsions. Bonding courses are painted, rolled or sprayed onto a clean undercourse.

Levelling course

If layers of components turn out to be uneven or rough, a levelling course may be needed. This compensates for component tolerances, creates a smooth surface and is supplied as bitumen roofing sheet, glass or plastic fleece, and as foam mats. It is laid loose or spot-glued.

Separating course

Separating layers or courses are laid to ensure that adjacent layers of structural elements with different expansion properties can move in relation to each other, or to accommodate other mechanical movements. They can also be used if two materials are chemically incompatible. The same materials can be used as for the levelling courses.

Vapour barrier

Vapour barriers are used to regulate moisture transmission inside the building. › see chapter Pitched roofs, Types of finish They are not waterproof, but simply inhibit vapour diffusion. The inhibition factor indicates how much moisture can diffuse through the vapour barrier. They can be bitumen roofing sheets, plastic vapour inhibiting sheets, elastomer roofing sheets, or compound foils. They can be laid loose or spot-glued. Points at which the individual sheets meet must be fully glued. The vapour barrier must extend to the top edge of the insulating course and be fixed there at roof edges and at penetrations. Vapour barriers can also be used to ensure that a roof is airtight.

Thermal insulation

Thermal insulation protects the building from heat loss in summer, and prevents large thermal build-ups through insolation in summer (summer thermal insulation). Expanded polystyrene (EPS), extruded polystyrene foam (XPS), extruded polyurethane (PUR), mineral fibre insulating material (MF), foam glass (FG), cork, wood fibre insulation material, or expanded

> \\ Hint:
> All roof areas must be built with a slope of at least 2% to the roof outlets. If the roof structure itself does not slope, a slope can be created with sloping screed (laid on the roof slab with a separating course) or with sloping insulation. The key feature in deciding between the two possibilities, after the question of expense, is whether the structure can take the additional weight of the screed or whether the insulation is available in appropriate thicknesses.

>
> \\ Tip:
> If the slope is created with insulation wedges, the individual insulated areas meet at an angle of 45° for rectangular floor plans. A valley is created at the joint. For the valley to have a 2% slope, the insulation wedges must have a 3% slope.

bituminized mineral infill may be used. If the thermal insulation is laid above the waterproofing course, care should be taken that all the subsequent courses be permeable to vapour diffusion, so that any moisture can escape. The thermal insulation can also be used to create a roof drainage incline. These insulation materials are supplied in an appropriate form by the manufacturer and are known as gradient insulation. Single elements are called insulation wedges.

Sheet insulation should be laid with an offset and joints that are as tight as possible. The insulation is glued to the surface below and the sheets are attached to each other according to the manufacturer's instructions.

Vapour pressure compensation course

A vapour pressure compensation course is fitted to distribute the water vapour pressure evenly in the roof waterproofing material. Such layers, like separating courses, may consist of bitumen roof sheeting, plastic vapour barrier sheeting, elastomer roof sheeting or compound foils. They are laid loose or spot-glued.

Roof waterproofing

The water-bearing course is created by the roof waterproofing. It is a closed, waterproof area and can be produced with bitumen roof sheeting, plastic or elastomer roof sheeting or fluid-applied waterproofing. Bitumen sheet waterproofing should have at least two layers. It must be laid with an overlap of at least 80 mm and be glued along the full area of the joints. Plastic or elastomer sheeting is applied in one layer. It must be glued over the full area. The overlap must be at least 40 mm. Fluid-applied waterproofing is laid in one layer. It must be applied so that it adheres over the full area. A separating course should be used if the course underneath is

Fig.37:
Flat roof coverings

open, for example for sheet insulation or a timber frame. The waterproofing must be at least 1.5 mm thick, at least 2 mm for used roofs.

Filter course

A protective or filter course is used to protect the waterproofing from mechanical influences. PVC, rubber or plastic granulate highly perforation-proof sheets, drainage mats or sheets of extruded polystyrene foam are used.

Surface protection

Modern foil roofs are installed as single-layer waterproofing courses and attached to the course below with strips that are glued on subsequently, or spot-attached. They are usually made of light-coloured materials, so that thermal expansion as a result of sunlight is minimized. Because of their material properties they offer adequate protection from UV radiation. No additional surface protection need be used for foil roofs. This saves weight, and the roof structure can be designed with less self-weight. Surface protection is applied if the roof waterproofing does not provide adequate protection against UV radiation, wind upthrust or mechanical loads. A light surface protection is provided for bitumen sheets according to load. This can be achieved by applying sand, for example. Crushed slate is sprinkled into a cold polymer bitumen compound and attached to the bitumen sheeting.

Gravel surface protection courses are known as heavy surface protection. The gravel layer should be at least 50 mm thick when installed. The weight of the stones can prevent unfixed roof coverings being lifted by wind suction. If the surface is protected with crushed stone or gravel, care should be taken that the grain size be sufficient for the material not to blow away in the wind. A sheet covering is preferable for surfaces exposed to strong winds.

Accessible coverings

Accessible coverings, i.e. coverings that can be walked on, can also be used to protect the surface. In such cases a compression-resistant thermal

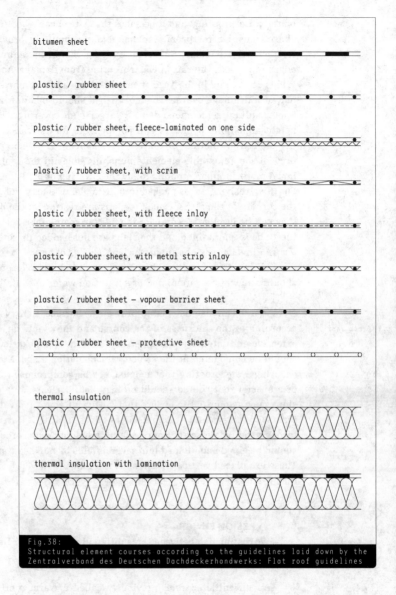

Fig.38:
Structural element courses according to the guidelines laid down by the Zentralverband des Deutschen Dachdeckerhandwerks: Flat roof guidelines

insulation course should be installed. The roof waterproofing must be adequately protected against mechanical influences.

If the joints between the slabs are sealed, or the entire covering is closed, a slope of at least 1% must be created. Drainage is then maintained on the waterproofing, and it becomes the water-bearing course. The slabs chosen must be frostproof. Expansion joints should be provided to allow for thermal expansion. Sufficient spacing should be maintained at the edges to prevent the raised waterproofing from damage. Accessible coverings can be laid in the form of small slabs in a bed of mortar on a tufted mat or a drainage course. The mortar bed should be approx. 4 cm thick. Spot bedding is recommended for larger slab coverings. Prefabricated height-adjustable elements can compensate for tolerances here. Compression-proof thermal insulation, such as foam glass, should be used to prevent the adjustable elements punching holes in the roof waterproofing. A more simple method is to lay the covering on little bags of mortar. For this, fresh mortar is packed into little plastic sacks and placed under the corners of the slabs. Once the slabs are resting on the mortar sacks, they can be levelled to compact the mortar completely. The cavity under the slabs is maintained and the slabs are not bedded directly on the waterproofing. Large slabs should be laid in a bed of gravel for better weight distribution. Here the gravel bed is approx. 5 cm thick. The gravel selected should allow any accumulated precipitation water to drain off freely.

Green roof

Plants can also be used to protect the surface. Here we distinguish, according to the thickness of the course and the plants chosen, between extensive and intensive planting. The additional load exerted on the roof should be considered at the structural calculation stage. Note also when choosing waterproofing that it must not be compromised by the roots, or a special root course should be installed. For extensive planting, the slope may be omitted to ensure that sufficient water is available for the plants. But care must be taken if damage does occur that water cannot seep through the whole set of courses. The individual waterproofing courses should be glued and divided into several fields by bulkhead-style barriers. These are placed vertically, and split the roof into several areas that are then drained separately. <u>Leaks are easier to locate</u>.

TYPES OF FINISH
We distinguish between three different principles for flat roof structures or course layers.

Unventilated roof

An unventilated roof (previously called a warm roof) has its waterproofing course on the outside and so the thermal insulation is in the sealed "warm area". A typical structure for an unventilated roof involves applying a preparatory coating to prepare the surface of the roof structure, which may be in reinforced concrete roof, steel or wood. The levelling course and the vapour barrier are laid on this surface. The insulation is

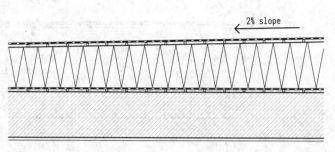

roof waterproofing, single course
vapour pressure compensation course
thermal insulation in sloping insulation slabs
vapour barrier
levelling course
preliminary coating
reinforced concrete slab

Fig.39:
Unventilated roof

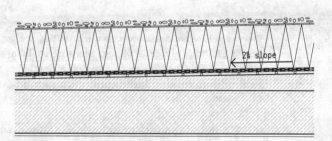

surface protection, gravel, min. 5 cm, grain 16–32
filter course
thermal insulation, durable, water resistant
waterproofing courses, 3 layers
levelling course
sloping screed
reinforced concrete slab

Fig.40:
Upside-down roof

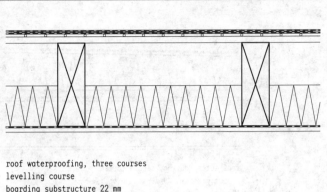

Fig.41:
Ventilated roof

roof waterproofing, three courses
levelling course
boarding substructure 22 mm
air space, min. 15 cm
thermal insulation
rafters
vapour barrier
plasterboard sheet

made of hard-wearing sloping slabs. Here great care must be taken that all areas of the roof drain towards a roof outlet or a gutter, with a slope of at least 2% (3% is better). A vapour pressure compensation course is laid on the insulation, and the waterproofing is applied to this; it can consist of one or more courses. Surface protection should be provided according to the product chosen.

Upside-down roof

The second approach to building up a layer structure is the upside-down roof. It is also called an IRMA roof (Insulated Roof Membrane Assembly). Here the thermal insulation course is above the waterproofing course, and must therefore be made of a water-resistant insulation material. For this structure, a sloping screed with an incline of at least 2% is placed on the roof support structure within the sequence of courses. The roof waterproofing is supported by a levelling course drained by the slope of the screed. The thermal insulation is also installed in the form of flat slabs. A filter course is placed on top to prevent elements of the surface protection material being washed into the insulation.

Ventilated roof

The third construction available is the ventilated roof (formerly also known as a cold roof). Ventilated roofs are often used for timber roof structures. The course structure is such that boarding, e.g. a sheet of plasterboard or chipboard, tops the floor below. A vapour barrier runs under the rafters above it. The thermal insulation is between the rafters, and may consist of a single layer. An air space of at least 15 cm must be left above the thermal insulation, between the rafters, to guarantee adequate through ventilation. Boarding is fitted on top of the rafters; this can consist of chipboard, tongue-and-groove boarding or a similar material. The waterproofing layers are laid on a levelling course. Surfaces can be protected where necessary.

FLASHING

Rising structural elements

To prevent spray or water that has accumulated on the roof from penetrating the structure the waterproofing must be taken higher. Structural elements rising from the roof may include higher sections of the building, lift headgear, chimneys or service spaces. The same waterproofing principles apply to windows and doors. Given a flat roof pitch of up to 5° the waterproofing must be continued and secured at least 15 cm above the top edge of the roof covering on the rising structural element. The top edge of the roof covering is not the waterproofing course, but may be the gravel surface protection. If the roof pitch is greater than 5° the waterproofing must be taken at least 10 cm up the rising structural element.

Door thresholds

Balcony and terrace doors present a particular difficulty. If the waterproofing is taken 15 cm above the working surface of the roof, there will inevitably be a step between the interior and the exterior. In most cases the shell height of the floor is the same inside and out, but the roof structure is much higher than the floor structure inside, because of the high proportion of waterproofing material. This requires an additional

> \\ Hint:
> Walls or columns standing vertically on the roof are examples of rising structural elements. The roof waterproofing meets a "rising structural element" at these points and must be appropriately secured to ensure that precipitation, spray or condensation water cannot run under the waterproofing course.

offset. But in order to make the roof area easily accessible, the flashing height can be reduced to 5 cm. This should ensure that if water accumulates in snowy conditions it cannot penetrate behind the waterproofing. If there is no roof outlet immediately outside the door, a grating or a gutter should be positioned there. Doors without thresholds may be essential when "barrier-free building" is required, e.g. in public buildings. This involves special construction methods, such as protection against spray by canopies, heated gutters connected directly to the drainage system, or a roof structure with fully bonded courses.

The waterproofing, and where applicable the separating course, can be secured with clamping rails, which are pinned to the rising wall. Plastic waterproofing can be glued to composite metal sheeting. Joints should also be inserted in the rails at points where the building has expansion joints, to avoid tension. The upward run of waterproofing can be masked with canted metal sheets attached to the rails to cover the waterproofing. These extend into the surface covering area. Care should be taken here that the sheet metal not cause mechanical damage to the waterproofing. For curtain façades, the waterproofing should be taken behind the façade, although it is essential that the waterproofing be easily accessible in case of damage.

All components should satisfy the appropriate fireproofing criteria. Care should also be taken in the case of rising parts of the building that a fire could not spread to higher sections through apertures in the roof, such as light cupolas. A gap of 5 m should be maintained to prevent possible flashover.

ROOF EDGING

Projecting flat roof

Flat roofs can be finished as projecting flat roofs at the point where they meet the façade, or with an upstand. Current practice is to finish the roof with a parapet or an edge trim. Projecting flat roofs present structural problems, as the waterproofing and insulating course in the roof and wall sections cannot easily be combined. If the external envelope is to be clad void-free in thermal insulation material, to avoid thermal bridges, the projecting section of the roof must be completely covered with thermal insulation material. This, however, makes it look unduly thick as a structural element. The second possibility is to apply the thermal insulation to the inside of the roof slab, which is not an ideal solution, as it is impossible to make a direct connection with the insulating course in the wall element. This would also place the roof slab structure in the cold outside area, so it would behave differently in relation to

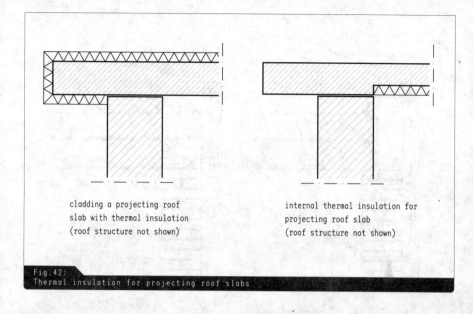

Fig.42:
Thermal insulation for projecting roof slabs

temperature changes from the loadbearing walls or columns below it. This could create tension at the connection points, which could cause cracks in the structure or the façade. This method is often used to provide existing listed buildings with thermal insulation subsequently, to maintain the appearance of the façade.

Roof parapet, upstand

For roof edging with upstands the waterproofing layers can be continued upwards, as for the rising structural elements in the flat roof structure. Here, the topmost point should stand at least 10 cm above the roof. The reference level is the topmost layer – the foil, gravel or working surface. The upstands can be finished as a roof parapet, e.g. in reinforced concrete or masonry, or with a roof edge trim mounted on an edging plank. The edging plank is usually a simple rectangular timber that can lie flat on the plane of the insulation. The roof edge trim is shaped to run around the edge of the roof and overlap part of the façade. If a roof parapet is chosen, metal sheeting is usually employed to perform this function, but there are also prefabricated stone and concrete versions. According to the height of the building, different dimensional recommendations apply, as the wind situation becomes more critical with increasing height. The parapet should project approx. 2 cm over the façade to create a drip edge and prevent any rainwater that may accumulate from running behind the sheet metal.

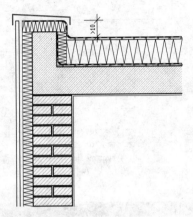
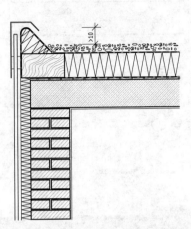

roof structure	wall structure	roof structure	wall structure
roof waterproofing, single course	façade slabs	gravel course	façade slabs
vapour pressure compensation course	air space	roof waterproofing, single course	air space
thermal insulation	thermal insulation	vapour pressure compensation course	thermal insulation
vapour barrier	calcium silicate masonry	thermal insulation	calcium silicate masonry
levelling course		vapour barrier	
preliminary coating		levelling course	
reinforced concrete slab		preliminary coating	
		reinforced concrete slab	

Fig.43:
Roof edge with parapet and edge trim

Table 8:
Façade overlaps for parapet sheets

Building height	Parapet sheet overlap
< 8 m	5 cm
8–20 m	8 cm
> 20 m	10 cm

DRAINAGE

Flat roofs are usually drained internally, i.e. the downpipes run through the building to the drainage system in the foundations. Every roof must have at least one outlet plus an emergency outlet. The dimensions of the outlet pipes are established as for pitched roofs. > see chapter Pitched roofs, Drainage

Slope

The roof should slope by at least 2% to take the water to the outlets, as it impossible to construct a completely flat roof. The slope prevents puddles from forming. It is created either with a sloping screed or sloping insulation. > see also section Insulation If the rooftop is to be accessible, accumulated precipitation water should be drained away on the surface as well as on the insulation plane. Roof outlets or roof gullies should be arranged so that they are at the lowest point on the roof, and freely accessible. It is possible to install inspection grids, for example, for used rooftops. These should be placed at least 30 cm away from rising structural elements or joints, so that the outlet can be cleanly waterproofed.

Roof outlets

The roof gullies must have a locking waterproofing ring for foils or an adhesive flange, bonded securely with the waterproofing course, for bituminous seals. Gullies for flat roofs are available with both vertical and angled inlets. The vertical version is preferable in principle, as the water is taken directly into the downpipe, and any leaks can be located quickly. Angled flat roof gullies are used when large connected areas are overbuilt, for example, and it is impossible to drain the water off directly (e.g. column-free spaces).

Heating the roof outlets is recommended in areas with heavy snowfall, so that they do not freeze. This ensures drainage in winter as well.

Emergency overflow

To prevent water from accumulating on the roof, when an outlet is blocked with leaves, for example, an emergency overflow must be installed. Emergency overflows are waterspouts running through the roof edging (parapet). They must be placed at a low point of the slope, and be waterproofed and insulated on all sides. The projecting section of the tube must be long enough to prevent water from running down the façade. The water does not have to be directed into the drainage system as an emergency overflow is not a permanent drainage feature.

Fig.44:
Emergency overflow

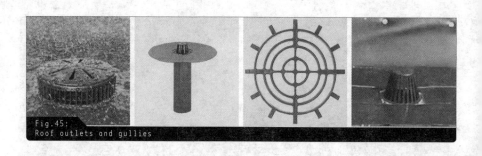

Fig.45:
Roof outlets and gullies

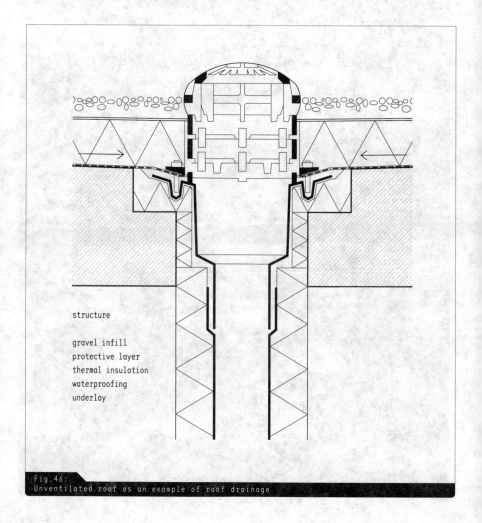

structure

gravel infill
protective layer
thermal insulation
waterproofing
underlay

Fig.46:
Unventilated roof as an example of roof drainage

IN CONCLUSION

The preceding chapters have introduced various roof forms, materials and designs. This book confines itself to simple forms, but it is soon clear that there is no such thing as a standard roof: so many different combinations are possible. Planners must define and fix the layer structure, along with all the connections, conclusions and penetrations. The following points should be borne in mind when working out the details:

Forms:
- Is the chosen roof structure compatible with the size and shape of the ground plan?
- Are access and a second escape route (e.g. through a window) guaranteed?
- Are the chosen roof form and, where applicable, the dormers admissible under the building regulations?
- Are the chosen materials suitable for the overall appearance of the building and its surroundings?
- Are lighting and ventilation for the roof space guaranteed?

Finish:
- Does the building have adequate thermal insulation on all its outside surfaces (gables, dormers, roof surfaces, roof superstructures, pipe runs)?
- Is the insulation joined to the insulation layer in adjacent structural elements at all points (walls, balconies etc.)?
- Is the roof reliably rainproof?
- Can moisture accumulating within a structural element escape?
- Can rainwater flow unhindered from all roof surfaces?
- Is the roof windproof?
- Is condensation (e.g. at penetration points) excluded?
- Have precautions been taken against atmospheric moisture penetrating the structure, and particularly the insulation?

Roofs offer great scope for design, in addition to all the conventional forms. New interpretations of familiar structural elements can be exciting features, while reduction to the essentials emphasizes strict order in the overall picture. Despite all the standards and regulations governing building, planners should develop a roof concept first; construction is the second step.

APPENDIX

STANDARDS

Loads and forces	
E DIN 1052	Entwurf, Berechnung und Bemessung von Holzbauwerken – Allgemeine Bemessungsregeln und Bemessungsregeln für den Hochbau, Berlin, Beuth-Verlag 2000 (Design, calculation and dimensioning for timber buildings)
E DIN 1055-1	Einwirkungen auf Tragwerke – Teil 1: Wichte und Flächenlasten von Baustoffen, Bauteilen und Lagerstoffen, Berlin, Beuth-Verlag 2000 (Effect on load-bearing systems, part 1, weights and area loads for materials, parts and stored material)
E DIN 1055-3	Einwirkungen auf Tragwerke – Teil 3: Eigen- und Nutzlasten für Hochbauten, Berlin, Beuth-Verlag 2000 (Work on load-bearing systems, part 3, own weight and imposed loads for building)
E DIN 1055-4	Einwirkungen auf Tragwerke – Teil 4: Windlasten, Berlin, Beuth-Verlag 2001 2001 (Effect on load-bearing systems, part 4, wind loads)
E DIN 1055-5	Einwirkungen auf Tragwerke - Teil 5: Schnee- und Eislasten, Berlin, Beuth-Verlag 2000 (Effect on load-bearing systems, part 5, snow and ice loads)
Sealing	
DIN 18195	Bauwerksabdichtungen, Teile 1-6 und 8-10, Ausgaben 8/83 bis 12/86, Berlin, Beuth-Verlag 1983/1986 (Sealing buildings)
Insulation	
DIN 4108	Beiblatt 2, Wärmeschutz und Energie-Einsparung in Gebäuden. Wärmebrücken. Planungs- und Ausführungsbeispiele (1998-08) (Supplementary sheet 2, heat insulation and energy saving in buildings, heat bridges, examples of planning and execution)
DIN 4108-2	Wärmeschutz und Energie-Einsparung in Gebäuden. Mindestanforderungen an den Wärmeschutz (2001-03) (Heat insulation and energy saving in buildings, minimum demands)

DIN 4108-3	Wärmeschutz und Energie-Einsparung in Gebäuden. Klimabedingter Feuchteschutz, Anforderungen, Berechnungsverfahren und Hinweise für die Planung und Ausführung (2001-07) (Heat insulation and energy saving in buildings, climate-related damp protection, requirements, calculation procedures and hints for planning and execution)
DIN V 4108-4	Wärmeschutz und Energie-Einsparung in Gebäuden. Wärme- und feuchteschutztechnische Bemessungswerte (2002-02) (Heat insulation and energy saving in buildings, heat and damp protection technical dimension values)
DIN 4108-7	Wärmeschutz und Energie-Einsparung in Gebäuden. Luftdichtheit von Gebäuden. Anforderungen, Planungs- und Ausführungsempfehlungen sowie -beispiele (2001-08) (Heat insulation and energy saving in buildings, air-tightness of buildings, requirements, planning and execution recommendations and examples)
SN EN ISO 10211-1	Wärmebrücken im Hochbau – Berechnung der Wärmeströme und Oberflächentemperaturen – Teil 1: Allgemeine Verfahren (ISO 10211-1:1995), 1995 (Heat bridges – calculating heat currents and surface temperatures, part 1, general procedures)
SN EN ISO 10211-2	Wärmebrücken im Hochbau - Berechnung der Wärmeströme und Oberflächentemperaturen - Teil 2: Linienförmige Wärmebrücken (ISO 10211-2:2001), 2001 (Heat bridges – calculating heat currents and surface temperatures, part 2, linear heat bridges)

Drainage

DIN 18460	Regenfallleitungen ausserhalb von Gebäuden und Dachrinnen (Rainfall drainage outside buildings and gutters)
DIN EN 612	Hängedachrinnen, Regenfallrohre ausserhalb von Gebäuden und Zubehörteile aus Metall, Berlin, Beuth-Verlag, 1996 (Suspended roof gutters, drainpipes outside buildings and metal components)
SN EN 612	Hängedachrinnen mit Aussteifung der Rinnenvorderseite und Regenrohre aus Metallblech mit Nahtverbindungen, 2005 (Suspended gutters with reinforced gutter fronts and sheet metal drainpipes with seam joints)
SSIV-10SN EN 12056-3	Schwerkraftentwässerungsanlagen innerhalb von Gebäuden – Teil 3: Dachentwässerung, Planung und Bemessung, Ausgabe 2000 (Gravity drainage inside buildings, part 3, roof drainage, planning and dimensioning)

LITERATURE

Francis D.K. Ching: *Building Construction illustrated*, 3rd edition, John Wiley & Sons, 2004

Andrea Deplazes (ed.): *Constructing Architecture*, Birkhäuser Publishers, Basel 2005

Thomas Herzog, Michael Volz, Julius Natterer, Wolfgang Winter, Roland Schweizer: *Timber Construction Manual*, Birkhäuser Publishers, Basel 2004

Ernst Neufert, Peter Neufert: *Architects' Data*, 3rd edition, Blackwell Science, UK USA Australia 2004

Eberhard Schunck, Hans Jochen Oster, Kurt Kiessl, Rainer Barthel: *Roof Construction Manual*, Birkhäuser Publishers, Basel 2003

Andrew Watts: *Modern Construction Roofs*, Springer Wien New York 2006

PICTURE CREDITS

Figure 3 left, figure 33 centre left (residential house in Wetter, Germany): Tanja Zagromski, Wetter, Germany

Figure 3 centre left, center right; figure 23 centre left; figure 25 centre left, centre right, figure 26 right; figure 37 left, centre right, figure page 16, 54: Bert Bielefeld, Dortmund, Germany

Figure 7 right (residential house in Oldenburg, Germany), figure 33 centre right, right (residential house in Lohne, Germany): Alfred Meistermann, Berlin

Figure Seite 40 (residential house in Naels, Germany): Brüning Architekten, Essen, Germany

Figure 33 left (Klenke Apartment Building); figure 44 (Villa 57): Archifactory.de, Bochum, Germany

All other figures are supplied by the author.

导言

屋顶作为建筑外表面的一个重要部分，需要满足一系列的功能需求：首先，屋顶需要保护其以下部分的空间，根据天气的情况进行开启或者关闭，最重要的是能够有效地排水、抵抗阳光和风荷载的作用以及提供私密的空间。

根据功能需求或者设计方法的不同，屋顶可以采用不同的结构形式。本书介绍了屋顶的一些最基本的概念和原则，而这些概念和原则为不断发展和涌现出的新的屋顶形式的设计提供了基础。

屋顶结构承受着不同形式的荷载，这些荷载必须直接地或者间接地（通过外墙、柱子或者基础）传递到地面上。

我们需要了解不同的结构形式和屋顶形式。在确定屋顶形式的时候，需要考虑一系列因素。其中外观可能是其中最重要的一个因素，然后是确定平面的布局和尺寸，而造价和相关的建筑规章也将起到至关重要的影响。

屋顶材料的选用应该与拟建项目的实际情况相符合：在私家住宅中一般很少采用精细加工的预制钢结构安装，同时在工业建筑中一般也应该尽可能避免采用现场手工的细部加工。

屋顶类型常常显示出一定的区域性特点。阿尔卑斯地区建筑的屋顶常常是平缓而且具有较长悬挑的坡屋顶；北欧沿海地区建筑的屋顶

> **提示：**
> 一般情况下，在进行一个新的建筑设计时，总体规划中已经规定了屋顶的形状和坡度。如果新建建筑位于较繁华的区域，同时又没有总体规划的相关规定，那么在进行屋顶设计的时候应该遵循"与周围建筑协调一致"的原则。通常，当地建筑部门会告知新建建筑的所在地点是否需要遵守特殊的相关规定。

常常坡度较大，而且搁置于面对马路的山墙之上（参考"屋顶类型"一节）。另外，建筑功能的不同也造成了屋顶形状的不同。比如，室内网球场的屋顶常常是顺着球场长向起拱的拱形屋顶，而普通的活动厅屋顶常常采用平屋顶形式，以便在不同场合下使用。

屋顶形状也可以采用不同类型的组合，但这种做法往往会导致屋顶的细部相对复杂，同时也增加了屋顶渗漏的危险。为了避免这种危险，在实际工程中屋顶结构更多采用的仍然是相对简单的形式。

屋顶大体上可以分为平屋顶和坡屋顶两大类。一般来说，当屋面倾斜角度大于5°的时候可以认为是坡屋顶。平屋顶与坡屋顶在结构和功能上均存在显著的差异，本书在后文中将分别进行分析和说明。

荷载与力

建筑结构的静力分析需要对作用于结构上的力以及其产生的效应进行相关计算，其中包括结构的稳定性分析。根据牛顿法则：力＝质量×加速度。通常，力的大小无法通过直接的方式确定，而只能通过其产生的效应来间接确定。打个比方，当我们观察到某一物体具有加速度的时候，我们就可以确定此时该物体上作用有一个或者多个力。然而，建筑结构的静力分析本质上是一个力平衡问题：建筑结构的所有不同构件均应该处于静止的平衡状态；同时结构的内力也必须处于平衡状态，故结构的各个构件均应具有一定的承载能力。构件材料的弹性、强度、厚度、尺寸等决定了其承载能力的大小。

当物体在荷载的作用下被压缩，此时物体承受压力；当物体在荷载的作用下被拉伸，此时物体承受拉力。当一个物体承受不在同一作用线且方向相反的两个作用力时，物体将会发生扭转和弯曲。在建筑行业中，我们称该物体在力矩的作用下承受弯矩或者扭矩。将所有施加于某结构上的力叠加所得的总荷载，就是该结构需要承受和传力的总荷载。（见图1）

作用于结构构件上的力还存在方向上的区分：沿着构件方向称为纵向力，垂直构件方向的称为横向力。

表1：
荷载

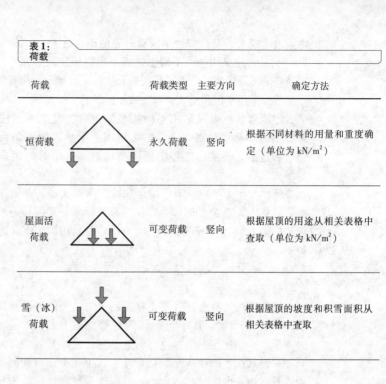

荷载	荷载类型	主要方向	确定方法
恒荷载	永久荷载	竖向	根据不同材料的用量和重度确定（单位为 kN/m^2）
屋面活荷载	可变荷载	竖向	根据屋顶的用途从相关表格中查取（单位为 kN/m^2）
雪（冰）荷载	可变荷载	竖向	根据屋顶的坡度和积雪面积从相关表格中查取
风荷载	可变荷载	方向可变	根据屋顶的坡度和受风面积从相关表格中查取

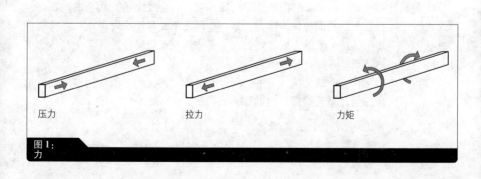

压力　　　　　拉力　　　　　力矩

图1：
力

建筑上作用有多种不同的荷载。在最初的设计阶段就需要确定荷载的大小以及荷载的传递方案。不同的荷载可能沿水平向（横向以及纵向）以及竖向作用在建筑之上，在确定屋顶尺寸之前需要首先确定作用在之上的每一种荷载的大小。设计者首先需要确定屋顶的选材，然后可以得到屋顶结构的自重。<u>恒荷载</u>是一种永久荷载，其作用方向为竖向向下。接下来需要考虑的因素是<u>屋面活荷载</u>，产生屋面活荷载的可能是一些可移动的物体，比如家具、人等。但在进行屋顶设计的时候并不需要列出每一件可能产生屋面活荷载的物体，而是采用根据建筑类型（比如住宅、工厂车间、仓库等）的不同采用相应的平均值，只有在非常特殊的情况下才需要单独考虑某一种物体产生的活荷载。即使人们通常不会到达屋顶结构的某个构件上，但我们在设计的时候仍然需要考虑维修和施工过程等因素，确保该构件能够承受一个人在上面行走所产生的荷载，该荷载为一个集中荷载。通常，屋面活荷载的作用方向与恒荷载相同，均为<u>竖向向下</u>。

　　风荷载、雪（冰）荷载均从屋顶外部作用到屋顶之上。雪（冰）荷载为施加在屋顶的压力，由于其产生于雪（冰）的自重，所以其作用方向也是竖向向下的。但风荷载可以同时从水平和竖向两个方向作用在屋顶之上。风荷载可以分为风吸力和风压力。其中风吸力将屋顶上提，所以在进行屋顶结构设计时需要采用必要的措施确保屋顶在承受风吸力的时候不被掀翻或者刮走。

注释：
　　各国的规范中均提供了荷载取值表格。结构构件的材料自重、雪荷载、风荷载以及屋面活荷载大小均可以从相应表格中查询。本书的附录中给出了一些重要的相关规范名称。

P13
坡屋顶

P13
基础知识

迄今为止，大多数的独立式住宅都采用了坡屋顶。坡屋顶在屋面排水方面具有平屋顶无可比拟的优势。虽然钢结构和混凝土结构都可以用作坡屋顶的承载结构，但目前更多地仍然是采用了手工制作的木结构。三角形桁架能够很好地将屋面承受的水平风荷载传递到建筑的主体结构之上。

坡屋顶的最高点称为<u>屋脊</u>，而较低的边称为<u>屋檐</u>。建筑<u>山墙</u>一侧屋顶的高、低点连线称为<u>山墙檐口</u>（见图2）。当两个屋面相交时，朝向建筑外部的交线称为<u>屋棱</u>，朝向建筑内部的交线称为<u>屋谷</u>。若搁置屋顶的墙体高度高于室内顶棚的高度，那么该墙体称为<u>胸墙</u>。<u>屋面倾斜角度</u>指的是屋面与水平面之间的夹角，一般可以采用两个平面之间的内角角度来衡量。但在进行屋面檐槽安装以及防水铺设的时候常用的则是屋面坡度，一般采用百分比的形式。

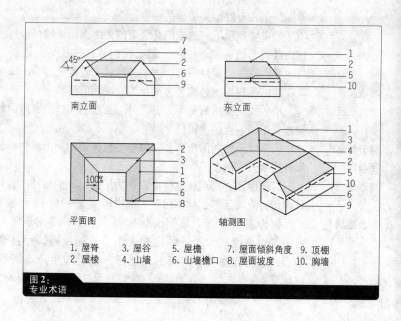

1. 屋脊　3. 屋谷　5. 屋檐　7. 屋面倾斜角度　9. 顶棚
2. 屋棱　4. 山墙　6. 山墙檐口　8. 屋面坡度　10. 胸墙

图2：
专业术语

图3：
单坡屋顶—人字屋顶—复折式屋顶—四坡屋顶

P14

屋顶类型

屋顶根据屋面类型和倾斜角度的不同可以划分为以下类型。（见图3，图4）

单坡屋顶　　单坡屋顶只有一个斜屋面，其檐口和屋脊位置搁置在不同高度的墙体之上。单坡屋顶适用于需要面对某一特殊方向的建筑中，比如面对花园的住宅以及面对街道的重要建筑等。

人字屋顶　　人字屋顶由两个相对放置的斜屋面构成。人字屋顶和单坡屋顶是坡屋顶最简单的两种形式。

复折式屋顶　　复折式屋顶的每侧由两个坡度不同的屋面相对放置构成，这种屋顶形式在目前使用得较少。当我们打算使用屋面下的空间时，采用这种屋顶形式可以增大房屋的净空。

如果坡屋顶下方的端墙是竖直的，那么该端墙可以称为山墙。若该山墙面朝街道或者广场，那么我们称该建筑为山墙面对街道；反之，我们称屋檐面对街道，而这种形式比较少见。

四坡屋顶　　四坡屋顶指的是四个方向均为斜屋面的屋顶。

攒尖式屋顶　　攒尖式屋顶指的是下方各边外墙长度相同，所有屋面均为坡屋面并相交于同一点的屋顶。

部分切割人字屋顶　　人字屋顶的两侧山墙顶部部分被坡屋面切割形成的屋顶形式称为部分切割人字屋顶。

筒形屋顶　　筒形屋顶是采用圆柱体曲面作为屋面形成的屋顶。若屋顶的每个方向均为曲面，那么称该屋顶为圆屋顶。

锯齿屋顶　　锯齿屋顶由若干按锯齿状排列的小型单坡屋顶或人字屋顶形成，其中坡度较大的一侧常常采用了玻璃屋面，而且全玻璃屋面的形式也比较常见，其目的是便于一些类似生产车间等大空间更好地采光。

图4：
攒尖式屋顶—部分切割人字屋顶—筒形屋顶—锯齿屋顶

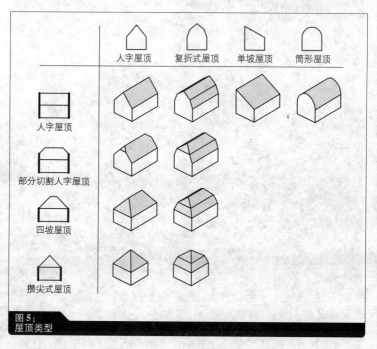

图5：
屋顶类型

提示：
"净空"是指可以被人较舒适地使用的空间高度。一般认为当净空高度大于2.2m时空间可以完全使用。

提示：
在计算顶层建筑面积的时候，净空在1~2m之间的部分仅计入一半的面积，净空大于2m的部分才全部计入，而净空小于1m的部分则不计入。各个国家和地区的相关规范对建筑面积的计算方法进行了相关规定。

有时需要对前文中的几个术语进行组合才能完整地描述一个屋顶的类型，比如部分切割复折式屋顶。对于只有一扇山墙的屋顶，并没有划分为其他的类型，直接划为人字屋顶、复折式屋顶或者单坡屋顶即可。

P17
屋顶窗

屋顶窗也是一种常见的屋顶类型。屋顶窗的作用包括便于屋顶采光、丰富立面造型以及增加顶层使用空间等。需要注意的是，屋面天窗以及屋顶窗的采光能力要远低于垂直墙面的窗户。所以对于采光而言，在山墙上设置窗户是一个更好的选择。屋顶窗的宽度需要控制在1~2倍的椽木间距之内（见"屋顶构造"一节）。屋顶窗可能还需要设置其他与房屋结构相连的承载构件。需要注意的是，屋顶窗的坡度不应低于规定的最小屋面坡度（见"屋面层"一节）。屋顶窗的所有面都必须满足和屋面层相同的密度、保湿、保温以及隔热等要求。

簸箕屋顶窗　　屋顶窗根据屋面层的不同而划分为不同的形式（见图6，图7，图8）。簸箕屋顶窗的屋面坡度小于主屋面的坡度，所以其两侧具有三角形的立面。

人字屋顶窗　　人字屋顶窗同样也具有竖向三角形的侧立面，与簸箕屋顶窗的不同之处在于人字屋顶窗与主屋面相交于屋谷，于是在屋顶窗上形成了面对街道的人字形屋顶。

三角形屋
顶窗　　当屋顶窗的山墙为三角形时，称该屋顶窗为三角形屋顶窗。

矮屋顶窗　　从屋面伸出具有较缓的弧线形状的屋顶称为矮屋顶窗。矮屋顶窗不会破坏主屋面完整性。其高度与簸箕屋顶窗相仿，高宽比通常在1∶10左右。

图6：
簸箕屋顶窗—人字屋顶窗—三角形屋顶窗

图7：
矮屋顶窗—舷窗—筒形屋顶窗

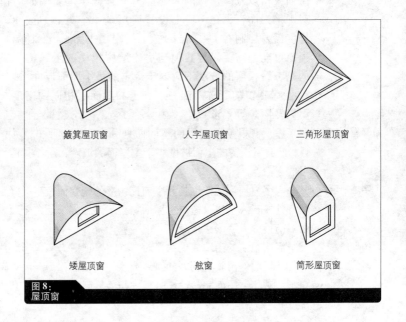

簸箕屋顶窗　　人字屋顶窗　　三角形屋顶窗

矮屋顶窗　　　舷窗　　　　筒形屋顶窗

图8：
屋顶窗

舷窗	舷窗在山墙的一侧呈现为半圆形。舷窗能够适用于各种不同类型屋面，但其上方的圆弧形区域需要采用金属层或者其他便于弯曲的材料来进行覆盖。
筒形屋顶窗	筒形屋顶窗与舷窗的区别在于其半圆形的部分下面还具有竖直的矩形部分。

图9：
承重木结构

P19

屋顶构造

和屋顶形式一样，我们也对屋顶构造进行了不同的分类。对于小型屋顶，比如私人住宅，木结构仍然是最好的选择。木结构能够较好地传递拉力和压力，同时还具有价格便宜和便于现场安装等优势（见图9）。钢结构或者预应力混凝土屋顶可能用在屋顶跨度较大的情况下，也可能用在住宅中以达到特殊的建筑效果。

坡屋顶中常用的承载体系主要有以下三种：对椽屋架、三角屋架和有檩屋架。

对椽屋架

对椽屋架是一个简单的三角框架体系（见图10，图11）。对椽屋架的横截面由两根上部倾斜相对的<u>屋椽</u>和底部的楼板或者<u>系梁</u>相连，形成了一个三角形框架。该三角形框架被称为"一对椽木"，其中构件之间确保连接点固定不同，但构件可以发生自由转动，也就是所谓的铰接。对椽屋架由若干对平行排列的椽木组成，每对椽木之间的距离通常在 70~80cm 之间，最大不超过 90cm。椽木主要承受屋顶的自重以及雪荷载等（见"荷载"一节）。系梁主要抵抗拉力，防止屋椽底部被拉开分离。因此，椽木与系梁或者顶棚之间的连接必须确保屋椽有效地将荷载传递到下部的墙体或者支承构件之上（见图12，图13）。在传统的屋椽工艺中，底部的系梁往往伸出至屋檐处，系梁伸出椽木的部分被称为"<u>边缘构件</u>"。

边缘构件

89

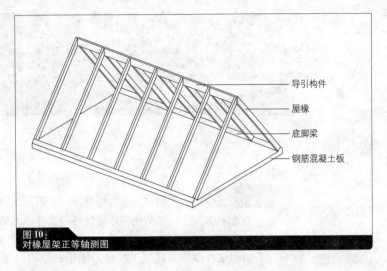

图10：
对椽屋架正等轴测图

导引构件
屋椽
底脚梁
钢筋混凝土板

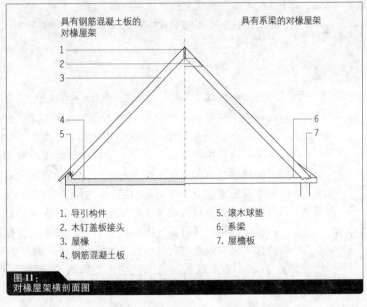

具有钢筋混凝土板的对椽屋架

具有系梁的对椽屋架

1. 导引构件
2. 木钉盖板接头
3. 屋椽
4. 钢筋混凝土板
5. 滚木球垫
6. 系梁
7. 屋檐板

图11：
对椽屋架横剖面图

| 屋檐板 | 边缘构件的长度通常应大于20cm，此时需要设置屋檐板将屋面延伸到边缘构件的外端，以确保屋面层能够覆盖至建筑外侧墙体。屋檐板的设置将改变屋面局部的倾斜角度，故需要对屋面层采用扭结或者铺设围边等措施。 |

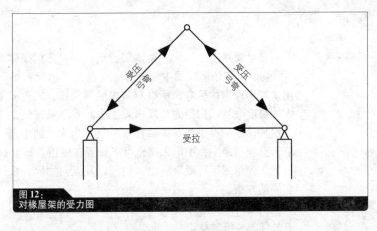

图 12：
对椽屋架的受力图

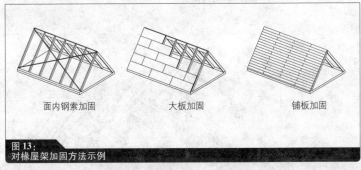

图 13：
对椽屋架加固方法示例

如今，实际工程的屋架中通常采用环梁的形式，其材料一般为混凝土。通过设置边梁或者在楼面设置锥面，帮助屋椽将斜向力传递到环梁中。

导引构件　　导引构件一般设置在屋椽对接的屋脊处，以便于屋椽的安装。导引构件从纵向将椽木连接在一起，通过在导引构件和椽木之间设计支撑木、金属条或者木板，可以确保屋顶能够有效地抵抗纵向荷载。

对椽屋架的倾斜角度大约在 25°～50° 之间，经济跨度在 8m 之内。当屋顶跨度大于 8m 的时候，所需构件截面较大，采用对椽屋架从经济角度来说不再是最优的选择，只有在无横墙平面设计的情况下可以采用对椽屋架，因为所有的屋顶荷载只能通过纵墙传递到地面。另外，只有需要在屋顶设置特殊的屋面构件（比如屋顶窗、大面积天窗和天窗承接梁）的时候，采用对椽屋架才是一个比较合适的选择（见"承接梁"一节）。

三角屋架

三角屋架

与对椽屋架相似,三角屋架也是一个三角框架体系(见图14,图15,图16)。三角屋架的中部设有一根称为"中间系梁"的水平系梁用来减小椽木的弓弯,所以与对椽屋架相比,三角屋架可以达到更大的跨度。从结构受力角度出发,中间系梁通常是在椽木的两侧成对水平设置的。中间系梁最佳水平位置为椽木的中点,但为了增加屋顶空间和净空,中间系梁也可以设置在屋顶总高度65%~70%的位置。三角屋架的屋脊、檐口构造以及纵向加固可采用与对椽屋架相同的方法。

当屋面倾斜角度大于45°、屋架跨度在10~15m时,三角屋架具有最佳的经济效益。

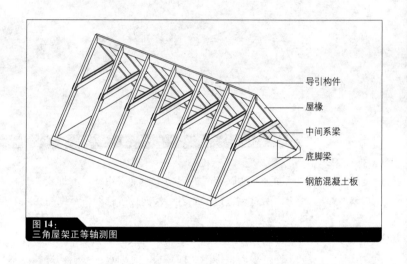

图14:
三角屋架正等轴测图

> **提示:**
> "跨度"指的是中部无支撑的结构构件长度。对于对椽屋架和三角屋架而言,跨度即为建筑的宽度。

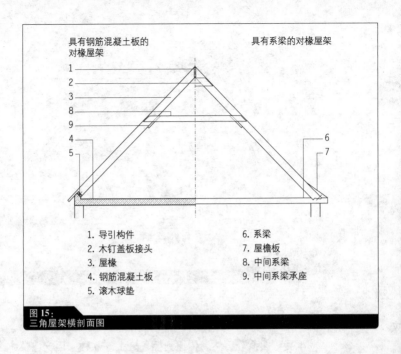

1. 导引构件
2. 木钉盖板接头
3. 屋椽
4. 钢筋混凝土板
5. 滚木球垫
6. 系梁
7. 屋檐板
8. 中间系梁
9. 中间系梁承座

图 15：
三角屋架横剖面图

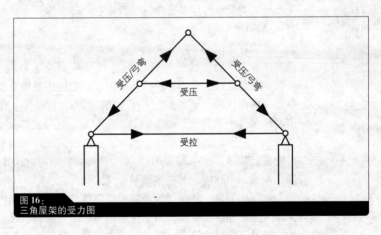

图 16：
三角屋架的受力图

有檩屋架

有檩屋架 和前两种屋架不同，有檩屋架具有水平承载构件——檩条——用以支撑屋椽（见图17）。檩条可以支撑在外墙或者<u>立柱</u>之上。屋椽用以

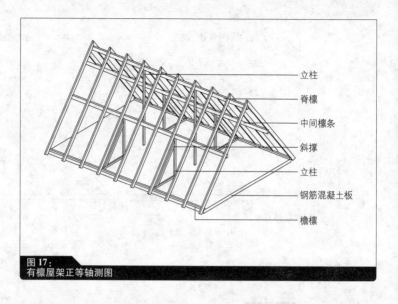

图17：
有檩屋架正等轴测图

立柱
斜撑

承受弯曲荷载，然后传递到檩条之上。立柱采用支撑拉牢之后可以承受水平风荷载。斜撑平行于屋椽设置在底部支撑和立柱侧面之间，使屋顶结构能够更好地承受横向荷载。

　　有檩屋架的基本形式为单檩屋架（见图18，图19）。在单檩屋架中，椽木放置于脊檩（位于屋脊位置）和檐檩（位于檐口位置）之上，脊檩的荷载通过立柱进行传递。由于屋架中仅在屋脊中设置了立柱支撑，所以称这种屋架形式为单檩屋架。在双檩屋架中，椽木由檐檩和中间檩条（一般设置在屋架的1/2高度处）进行支撑。由于椽木的跨度得到了降低，其挠度减小。同时，可以通过设置中间系梁对屋架的横向进行加强。如果屋顶的平面尺寸较大，则可以采用同时具有脊檩、檐檩和中间檩条的三檩屋架形式。

　　有檩屋架是一种功能性和通用性最强的经典屋顶结构形式。由于椽木之间是相互独立的，便于建造出多种不规则和混合形式的屋架结构。通过承接梁的设置，可以为烟囱以及天窗等构件的设置提供很大的方便。（见"承接梁"一节）

　　有檩屋架的坡度可以按照需要任意选择，常用的檩条间距可以达到4.5m。

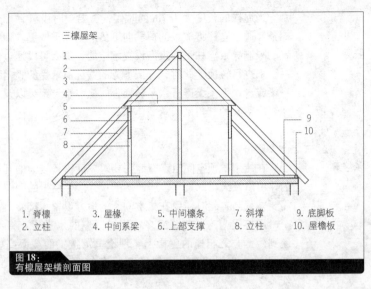

图 18：
有檩屋架横剖面图

1. 脊檩
2. 立柱
3. 屋椽
4. 中间系梁
5. 中间檩条
6. 上部支撑
7. 斜撑
8. 立柱
9. 底脚板
10. 屋檐板

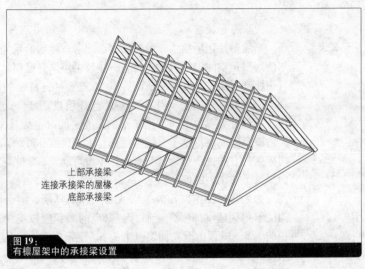

上部承接梁
连接承接梁的屋椽
底部承接梁

图 19：
有檩屋架中的承接梁设置

对于有檩屋架而言，非常重要的一点是立柱所承受的荷载必须有效地传递到房屋的承载体系中。立柱的下方必须设有承重墙、托梁或者其他竖向承载构件之上。

托架

当屋椽的长度大于7m的时候,需要同时设有脊檩、檐檩和中间檩条,通过立柱将檩条的荷载向下传递到承重墙上。如果建筑平面布置的限制导致立柱下面没有设置承重墙,那么可以通过设置托架或者系梁来进行转换。此时,脊檩下方的立柱所承受的竖向荷载经过中间系梁或者水平系梁进行水平转换,屋椽、水平系梁以及立柱形成了类似对椽屋架的三角框架承载体系。

承接梁

承接梁

在使用屋顶空间时,有时需要对屋面进行较大的开洞,比如烟囱的伸出以及天窗的设置。当开洞的尺寸大于屋椽的间距时,部分椽木需要被截断,此时需要在椽木间断的地方设置承接梁。

承接梁为跨越一跨或多跨屋椽区域的水平梁,将承受的荷载传递到两侧与之相连的屋椽上。理论上承接梁的截面高度应该与屋椽相同,如此可以确保屋架平面厚度一致,以便于其他结构构件的安装。

采光屋顶

天窗与屋顶窗

当需要使用屋顶空间的时候,必须考虑屋顶的采光和通风。如果不使用屋顶空间但可以供人行走,建议采用采光屋顶。采光屋顶的设置需要和屋面层的采用相适应,并不需要进行特殊的防风和隔热处理,但采光屋顶的造价往往较高而且必须便于维修。采光屋顶的使用,不仅可为室内空间提供基本的采光,同时也为从室内检查屋顶、天沟以及屋面层提供了光线和便利。由于建筑一般较高,采用爬梯通向屋面的方式过于危险甚至有时候不太可能,而采光屋顶则为通向屋面提供了非常大的便利。

如果将屋顶空间作为生活或者工作的场所,那么必须保证该空间能够得到适度的采光。通常,窗户的面积应该不少于房屋地基面积的1/8。屋顶山墙立面窗(若有)、屋顶窗和天窗均可为屋顶提供采光和通风。山墙窗和屋顶窗由于均为竖向,雨雪等不太容易直接附着到玻璃之上,相对比较耐脏;同时,人们可以非常容易地来到窗前欣赏窗外的美景。从防火逃生角度出发,屋面天窗可以作为"第二防火通道"(第一安全通道为下楼楼梯)。所以,天窗的尺寸必须足够大,当火灾发生的时候位于屋顶空间的人才有可能注意到头顶的天窗。

> **提示：**
> 相关法律法规给定了将屋面天窗作为逃生通道的时候的基本要求。一般情况下认为 90cm × 120cm 的窗户已经足够大；胸墙的高度不应高于 120cm，以确保能够从室内较容易地逃到窗外；同时天窗的位置离屋檐的距离不应大于 120cm，否则就需要在窗下设置救生平台以吸引人们的注意力。

对于屋顶空间的通风，在设计阶段就应该考虑到屋顶空间的气温很容易上升。如果顶层使用到了屋脊附近的空间，必须采用必要的措施确保空气可以从屋顶的上部空间中排出。

屋面天窗

如今，屋面天窗一般由屋面承包商作为预制构件提前制作完成，同时有多种能够适应不同屋面层的连接装置和配件可以选用。预制天窗的尺寸可以满足常见的屋椽间距（一般为 70~90cm）。屋顶天窗不仅可以安装在两榀椽木之间或者其间距之内，也可以跨越几榀椽木的距离，不过此时需要设置承接梁（见"承接梁"一节）。如果需要成排设置天窗，则可以采用组合框架取代天窗上、下侧的承接梁以便于天窗成排安装。天窗的高度应严格控制在约 1.6m 之内，否则其自重过大，在开启或者关闭的时候将存在一定的困难。如果需要安装窗户的区域较大，可以选择同时安装两扇天窗。天窗的上边缘宽度至少不应该小于 1.9m，以便于人向外观看。常见的胸墙高度在 0.85~1m 之间，对于厨房或者浴室胸墙高度可以适当提高。屋面坡度越小，达到相同高度所需的天窗面积越大。在进行天窗设计时，需要避免天窗破坏屋顶空间的空气流通，确保空气能够在天窗周围自由循环。天窗的四周均应该设置护窗板（比如，可以在天窗上侧打入楔子），确保水可以无障碍地快速流过。屋面的保温材料必须一直设置到窗边，而且天窗与屋顶之间的间隙也必须填满保温材料。天窗的周围还必须设置蒸汽隔离层，以避免室内空气中的水蒸气不会渗透到屋面保温材料中。

屋面天窗根据其不同的打开方式存在不同的分类。<u>中悬窗</u>固定在两侧的中点处，当开启的时候中悬窗的上半部分向内旋转到屋顶空间，下半部分向外旋转到室外空间。<u>上悬窗</u>则是固定在窗户的上边缘，可以将整个窗户向外或者向内开启。由于悬窗外侧的清洁比较困难，所以使用较多的是<u>折叠式中悬窗</u>。另外，有一些上悬窗也同时具有平推功能，将上悬窗向外推开之后可以平推到屋面之上，这种窗户的制作工艺相对比较复杂，同时造价也较高，目前使用得较少。

在布置天窗的位置时需要确保窗户之间、窗户与相邻建筑或其他边界之间具有足够空间。相关建筑规范规定天窗距防火墙的距离必须大于1.25m，而对于山墙面对街道的排屋（见"屋顶类型"一节）天窗离屋檐的距离不应小于2m。对于规范不允许设置屋顶窗的房屋常常采用天窗作为替代。

屋顶窗

和天窗相比，采用屋顶窗进行采光可以增加净空，同时提供更多的使用空间。屋顶窗内部的高度不应小于2m。和屋顶天窗相同，屋顶窗下部的胸墙高度也在0.85~1m之间。小型屋顶窗可以设置在两跨椽木之间，其宽度与椽木间距相同（大约在70~80cm之间）。当屋顶窗宽度增大时，需要设置承接梁。对于对椽屋架，屋顶窗的最大宽度不应大于椽木间距的2倍。对于有檩屋架，屋顶窗的宽度可以更大，而椽木承受的荷载将竖向向下传递到顶棚上。

屋顶窗的正立面由一个矩形框架组成，该框架称为<u>屋顶窗框架</u>，框架可以放置到椽木或者楼面之上。当屋顶窗框架放置在楼面上时，屋顶窗框架和檐口之间的空间通常是封闭起来不被使用的。对于单边屋顶窗，其窗顶主椽木的后端可以直接搭接在屋顶对应位置的椽木上，前端由屋顶窗框架支撑。两侧的三角形立面称为<u>屋顶窗侧墙</u>。屋顶窗侧墙可以选用粗糙的舌榫板或者其他平板，该区域也应该和屋顶的其他区域一样进行密封和保温隔热处理。对于屋顶窗侧墙，通常其外侧覆盖一层适用于三角形区域的金属板或者石材。另外也可以在屋顶窗侧墙安装玻璃，这样可以相应增加屋顶窗的采光量。

屋顶窗的宽度根据其覆盖范围内的屋面材料属性来确定，常常为屋面波形瓦（或者其他屋面覆盖材料）宽度的整数倍。屋顶窗的长度确定方法与其宽度类似（见"屋面木条"一节）。

如果工期比较紧张，那么建议采用预制屋顶窗。预制屋顶窗的安装一般可以在一天之内完成，而且可以在最短的时间之内恢复屋面的防水功能。

屋顶结构分层

屋面覆层

屋面覆层的主要功能是确保雨雪能够及时有效地排出，比如避免在大风雪情况下湿气渗入到室内。屋面覆层应该具有防风、防雨功能以及防火功能。另外，屋面覆层必须确保室内的潮气能够排到室外，同时确保其覆盖的结构构件避免风荷载的侵袭。屋面覆层需要根据屋顶布局、屋面坡度和屋面形状等关键因素进行选择。采用小块的平嵌瓦（见"平瓦"部分内容）可以比较容易进行屋谷或者屋面角部的铺设，而对于屋面大面积的平整部分采用波形瓦则更加方便和经济。然而，不同的屋面覆盖材料往往对应着各自适用的最小屋面坡度。在没有特殊说明的情况下，制造商对不同屋面覆盖材料所规定的适用屋面坡度均为最小屋面坡度。如果屋面的局部区域无法到达所规定的坡度，则需要在覆层以下铺设其他材料（见"屋面防水"部分）以防止雨水和灰尘渗入到屋顶内部。市场上有各种不同类型和不同材料的屋面覆层可以选用（见图20~图24）。

茅草屋顶

如今仍然可以在少数一些地方看到和古代一样采用芦苇或者稻草覆盖屋面的做法。这种做法要求屋面的倾斜角度大约45°。当屋面倾斜角度为50°的时候，风荷载将对屋顶产生向下的压力，而向上的吸力较小，对于茅草屋顶是一个比较理想的倾斜角度。屋顶的茅草一般为成束重叠覆盖在屋顶框架之上。

平瓦

平瓦的材料可以是木头（木瓦）、石头、混凝土或者黏土。木瓦适用的屋面坡度和木瓦的长度、搭接尺寸以及层数相关。对于普通的双层木瓦，其需要的最小屋面倾斜角度达到了70°，也就是接近竖直。而对于精细的三层木瓦，所需的屋面倾斜角度则降到了22°。

平石瓦一般是由页岩加工而成，形状包括矩形、锐角三角形、圆贝形以及鱼鳞形等，并根据不同的形状采用不同的方式在搭接处将其固定到屋顶结构上。平石瓦所需的最小屋面倾斜角度大约在25°~30°。

混凝土瓦和黏土瓦均需要进行工业化生产。在屋面边缘以及与相邻屋面相接的地方适合铺设购买的混凝土瓦或者黏土瓦。如果瓦片需要铺设在屋面木条上，那么可以在其下方设置立柱。混凝土瓦和黏土瓦所需的最小屋面倾斜角度大约在25°~40°。

异形瓦　　异形瓦形状多样，其材料一般为混凝土或者黏土，和平瓦一样均需要进行工业化生产。其不同的形状可以满足多种特殊的需要。与平瓦不同，异形瓦在进行铺设的时候需要进行三边搭接。

凸筒瓦和凹瓦　　屋面瓦的最早形式为凸筒瓦和凹瓦（见图25）：空心圆锥形的瓦片上下交错放置形成互锁的状态。上部的瓦片凸起，将雨水分散到两侧下方的凹瓦中，然后凹瓦将雨水引入到天沟内。凸筒瓦和凹瓦没有边棱也不设肋。如今的凸筒瓦和凹瓦则是从瓦片的上下边缘进行互锁，以防止发生渗漏。不同异形瓦所需的屋面最小倾斜角度不同，大约在22°~40°。

大型异形瓦还有其他的类型（比如波纹瓦）和其他的材料，打个简单的比方——纤维水泥瓦。它们在屋面木条处进行搭接，其宽度可达1m左右。不同的制造商提供了不同的封边、交叉和支撑方法，可以在屋面坡度小于12°的情况下使用。在屋面坡度更低的情况下可以采用沥青波纹瓦，此时瓦面的封边和连接之处一般采用了金属薄板包角的做法。

> **提示：**
> 对于不同屋面覆层所对应的屋面最小倾斜角度，可参见《Roof Construction Manual-Pitched Roofs》一书，Eberhard Schunk et al., Birkhäuser.

图 20：茅草屋顶

图 21：平瓦：木瓦屋面

图 22：平瓦：页岩石瓦屋面

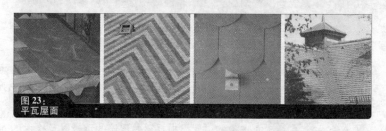
图 23：平瓦屋面

图 24：异形瓦屋面

工业建筑　　　有一些金属异形瓦甚至可以使用在屋面倾斜角度为5°的标准屋顶之上。这些异形瓦一般由涂有电镀层的钢板、铜板或者铝合金板制作而成，其截面形状和纤维水泥瓦、沥青波纹瓦、梯形断面压型钢板相似，为波纹形或者梯形（见图26）。梯形断面压型钢板由金属薄板压轧而成，具有多种不同的断面形状和尺寸，钢板的边缘进行了优化设计，以确保压型钢板具有更好的承载性能和更大的跨度。工业建筑中采用一般为组合形式的梯形断面压型钢板，其钢板之间夹设了保温隔热层。压型钢板之间必须进行搭接并相互可靠连接，以确保屋面的防水性和气密性。压型钢板通过螺钉、栓钉或者夹具与下部的支撑结构相连。在采用金属覆层的时候需要特别注意的一点是避免金属覆层与其他金属之间产生接触腐蚀。当屋面采用了和金属覆层不同材料的金属檩条或者混凝土檩条时，需要对屋面覆层和檩条之间进行隔离处理（见"平屋顶"一章中"屋面结构分层"部分的内容）。

金属板带　　　金属屋顶覆层的另一种形式是金属板带，包括铅板、铝板、铜板、不锈钢板或者电镀钢板等。金属板带的宽度一般在500～1500mm之间，可以按行或者按列放置。屋面两侧边缘处可采用接缝、卷拢或者搭接的方式进行封边处理，屋面的上下边缘处可以采用横向接缝或者搭接的方式进行封边处理。金属板带与其他构件相连的地方则可以采用人工卷边搭扣的方式。这种屋面覆层所需的屋面最小倾斜角度为5°，当屋面倾斜角度更小时则需要采用其他一些特殊的处理措施。

屋面木条

屋面覆盖材料可以通过螺丝钉、铁钉、螺栓或者夹具与屋面木条连接，而当屋面瓦设有支杆的时候则是由支杆与屋面木条相连。

屋面木条　　　屋面木条的尺寸需根据屋面覆层的重量以及椽木的间距选用。对于常见的屋面覆层，以下给出了屋面木条选用的建议方法：椽木间距小于30cm——屋面木条尺寸24/48mm；椽木间距小于80cm——屋面木条尺寸30/50mm；椽木间距小于100cm——屋面木条尺寸40/60mm。同时还需要考虑木条材料的强度等级。

图25：
凸筒瓦和凹瓦屋面

图26：
异形瓦屋面

椽木间距的确定

橡木间距的选择取决于屋面坡度和屋面覆盖物的性质，椽木的价格也根据其种类和生产商的不同而存在差异。在确定椽木间距的时候，首先需要确定屋顶覆盖区域的宽度，该宽度近似等于椽木的长度。屋顶的下部边缘处是一个比较特殊的区域，考虑屋顶建筑的整体性，设计者必须对该区域给予特别的注意，必须确定屋顶最后一排屋面瓦是否伸出屋椽之外、选择与屋面相平或者是向上倾斜的方式等。通常的方式是根据屋面瓦搭接的情况，让最后一排屋面瓦悬挑出屋椽。该悬挑长度称为<u>出檐长度</u>。在屋脊区域，搭接于屋面木条的屋面瓦需要隔开一定的间距以确保相互独立，该距离为<u>脊瓦间距</u>。在计算所需屋面木条的行数时，需要用屋面的总覆盖宽度减去出檐长度和脊瓦间距之后的长度。然后根据该长度与制造商所提供的屋面瓦覆盖长度相除得到所需的屋面木条的行数。

所需屋面瓦的数量同样需要根据制造商所提供的屋面瓦尺寸进行计算，而计算的数目同样会受到屋面烟囱、出屋面管道的细微影响。

图 27：屋面木条

表 2：屋面木条的截面选择

屋面木条截面尺寸（mm）	适用跨度（m）	木材强度等级（根据德国标准化学会《DIN 4074》规范）
24/48	≤0.70	S 13
24/60	≤0.80	S 13
30/50	≤0.80	S 10
40/60	≤1.00	S 10

当采用其他截面的屋面木条或者跨度时必须进行相应的静力测试。

屋面交叉木条

当屋面铺设的平瓦设置垫层或者屋面倾斜角度小于22°的时候需要设置屋面交叉木条（见图27，图28，表2），对于处于屋面排水不畅位置的中间楼面同样需要采用该措施（见"屋面防水"一节）。

实例：

屋面木条间距计算示例

屋面覆盖总宽度：7.08m

出檐距离：32cm（根据设计情况）

脊瓦间距：4cm（根据制造商的产品说明）

木条平分距离：7.08m − 0.32m − 0.04m = 6.72m

根据制造商的产品说明，屋面木条平均间距为0.33m

6.72m：0.33m = 20.4

确定的屋面木条行数：20 行

需要特别注意的是以上所选用的屋面木条平均间距0.33m 是否满足制造商对于待建屋面坡度所给出的适用间距要求。

注释：

并不是在设有屋面垫层的情况下一定要设置屋面交叉木条。而屋面垫层会向下产生一定的挠度，可能从屋面渗入的雨水将在屋面木条下方安全地排出。但此时仍然建议采用屋面交叉木条，因为随着时间增加，屋面垫层将发生收缩而使其处于过度受拉的状态。

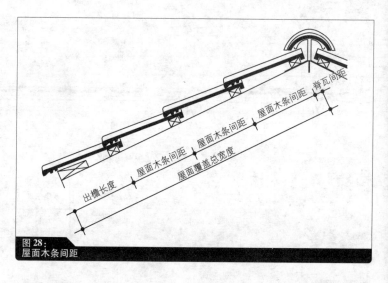

图28：
屋面木条间距

屋面防水

一般来说，屋面覆层可以确保屋面的防水功能。然而，针对大风、暴风雪等一些特殊极端天气所造成的喷射渗透需要采用其他额外的措施。过低的屋面坡度、过于结构化以及特殊的结构形式或者屋面活动空间都可能导致需要采用额外措施确保屋面防水。另外，特殊的气候条件（比如暴晒、大风和经常大雪的区域）也可能需要对屋面采用额外措施，确保屋面防水。

屋面垫层

最简单的额外措施是铺设屋面垫层（见图29，表3）。垫层的安装和薄板通风结构相似，也就是说，垫层并不需要进行下部支撑，而可以自由地悬挂在橡木之间。屋面垫层一般是成卷出售的，常见的形式是强化塑料膜。

有支撑屋面垫层

此时垫层下部设有支撑，比如木结构等。有支撑垫层也属于防水层，其防水效果取决于接缝的质量。有支撑垫层设置在屋面木条或者屋面交叉木条的下方。

熔接屋面垫层

将具有防水性能的塑料薄膜采用热熔或者胶合的方法连接就可以组成防水垫层，进一步可以细分为防雨垫层和防水垫层。对于防雨垫层，其中可能包含一些必需的结构开洞。防雨垫层布置在屋面木条或屋面交叉木条的下方。对于防水垫层，其中不允许进行任何开洞，设置于屋面木条与交叉木条之间，与屋面木条形成一个整体。在进行垫层安装的时候需要特别注意确保其上的雨水可以顺畅地排入到屋檐处的天沟中，同时也要确保垫层下方屋顶空间内具备适当的通风条件。

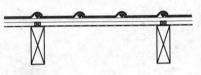

不设屋面交叉木条　　　　　　　　设屋面交叉木条

屋面覆层
屋面木条
屋面垫层（悬挂安装，便于通风）
椽木

屋面覆层
屋面木条
屋面交叉木条（便于垫层通风）
屋面垫层
椽木

图29：
屋面垫层铺设

表3：
屋面覆层所需的额外措施

屋面倾斜角度	由于功能、结构、气候等因素所需的额外措施			
	无特殊因素	一种特殊因素	两种特殊因素	三种特殊因素
大于所需最小屋面倾斜角度	—	屋面垫层	屋面垫层	有支撑屋面垫层
大于10°，小于最小屋面倾斜角度	屋面垫层	屋面垫层	有支撑屋面垫层	熔接屋面垫层
大于6°，小于最小屋面倾斜角度	防雨垫层	防雨垫层	防雨垫层	防水垫层
小于6°，同时小于最小屋面倾斜角度	防雨垫层	防水垫层	防水垫层	防水垫层

保温隔热层

如果屋顶空间具有通风条件，同时也不作为生活空间使用，那么保温隔热层一般可以设置在最上层的顶棚内（见图30）。和将保温隔热层设置在屋面中相比，这种做法既便于施工又可以减少保温材料的用量。然而，现在屋顶空间越来越多地被设计成生活场所以增加室内的使用空间。在这种情况下，屋顶空间将成为一个受热区域，包围该区域的所有结构构件都需要满足相关的保温隔热规范要求。（见"附录"）

非常重要的一点是确保屋面保温隔热层与外墙的保温隔热层相连，避免形成冷桥区域（见图31）。

保温隔热层

矿物棉、硬质泡沫塑料板（PS）、硬质聚氨酯泡沫板（PUR）、软木、轻质刨花板以及浇注成型的颗粒材料均可以当作屋面保温隔热材料使用（见图32）。不同的材料具有不同的热导率，热导率越低说明保温隔热性能越好，在效率相同的情况下采用热导率低的材料可以使保温隔热层越薄。热导系数（k）的单位为瓦特每平方米开尔文 [$W/(m^2 \cdot K)$]，该数值越低说明保温隔热性能越好。

保温隔热层可以设置在椽木之间，此时椽木必须有足够的高度来确保保温隔热层的铺设。在有些情况下为了满足保温隔热材料的安装要求，可能所选的椽木高度大于受力角度所需的高度，否则需要在屋顶内侧向椽木上另安装一层保温隔热材料。一般椽木的木材热导性能非常弱，即使保温层在椽木所在的位置间断也不会导致冷桥的产生。

提示：

冷桥区域指的是建筑中保温隔热层中断的地方。当建筑内外存在温差的时候，热量可以通过冷桥区域进行传递，当室内温度较高时（比如冬天）热量逃至室外，当室外温度较高时（比如夏天）热量进入室内。由于水汽可能在温度较低的表面凝结聚集，然后渗入建筑中对结构造成损害，故要尽可能避免出现冷桥区域。

注释：

不同材料的保温隔热层大都可以满足常用的椽木间距，所以在设计的最初阶段设计者就应该搜集相关的产品资料。建议在确定椽木间距的时候选择一个固定的满足保温隔热材料铺设的间距，而不是直接采用一个整数单位值。

图30：
坡屋顶保温隔热材料的铺设

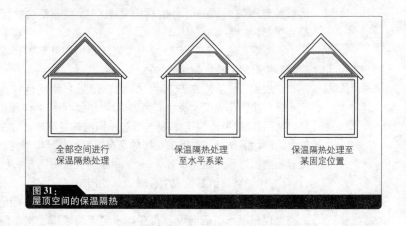

全部空间进行保温隔热处理　　保温隔热处理至水平系梁　　保温隔热处理至某固定位置

图31：
屋顶空间的保温隔热

屋椽间保温隔热层
全屋椽保温隔热层
屋椽上保温隔热层
蒸汽隔离层

屋椽间保温隔热层指的是在其和下部铺设层之间设有空隙以便于结构通风的做法。如果屋椽的高度与保温隔热层的厚度正好相等，则称为<u>全屋椽保温隔热层</u>。如果屋椽从室内可见，而保温隔热层设置于屋椽外侧的板材上，则称该保温隔热层为<u>屋椽上保温隔热层</u>。

对于设有保温隔热层的屋顶而言，设置蒸汽隔离层是一项必不可少的措施。蒸汽隔离层一般设置在屋顶空间的内侧、保温隔热层的下部，而且必须覆盖整个屋顶空间。蒸汽隔离层的边缘必须固定，以确保蒸汽隔离层的气密性。蒸汽隔离层的作用是防止室内空气中的水蒸气渗透到屋面保温隔热层或者屋面结构中。

蒸汽隔离层的绝缘率需要根据屋面坡度和椽木的长度确定，该值表示的是空气层的等效空间扩散深度（s_d），用以衡量材料抵抗水蒸气扩散的能力。蒸汽隔离层可以选用橡胶薄膜或者塑料薄膜。

> **提示：**
> 当室内空气通过保温隔热层从内向外扩散时，其温度也将随之降低（冬天）。由于热空气中的水蒸气含量大于相应的冷空气，所以在扩散的过程中将会产生水汽凝结而对屋面保温隔热层产生损害，进而对整个建筑产生损害。
> 如果房屋内某些部分设置了多层蒸汽隔离层，那么外层的材料应该比内层的材料更容易让水蒸气通过，从而使可能聚集在保温隔热层里面的水蒸气能够更加容易地扩散到空气中。

> **注释：**
> 只要能够确保内部空间的气密性，固体建筑材料也可以用作蒸汽隔离层。比如，采用 OSB 板（定向刨花板）封闭连接（使用合适的胶粘带）也可以到达封闭的效果，但此时一般需要测试以确认其气密性。

总之，必须确保屋顶结构、屋面保温隔热层（尤其重要）避免遭受水分渗透。需要通过内外两方面达到该目标：屋顶内部设置蒸汽隔离层防止空气中的水汽渗透；屋顶外部空气中的水汽由屋脊和屋面覆层遮挡，所以可能聚集的水通过所铺设的屋面覆层以及其他防雨（水）层导向天沟排出。

P39
不通风屋面结构

面层铺设方法

不通风屋面层结构一般采用全屋椽保温隔热层。屋面保温隔热层位于蒸汽隔离层（位于内侧）和屋面垫层（位于外侧）之间。屋面交叉木条铺设于屋面垫层上部，确保渗入到屋面覆层内部的湿气能在该空隙中得以挥发。当要求结构构件尽可能小的时候可以采用这种建造方法，另外若屋顶结构中后来增设了其他结构导致室内空间不足以设置通风屋面结构时，也可以采用这种屋面结构形式。但若此时屋面的通风对于保温隔热层非常重要必须设置的时候，可以采用在椽木外侧增设木条以增加椽木高度或者从屋顶内部另外增设一道保温隔热层的措施。

通风屋面结构	当采用了通风屋面结构时，屋面垫层下部的空气可以自由循环。其优点是屋顶结构中的湿气可以从屋脊处的通风口排出而不聚集到保温隔热层中。同时由于空气层的存在，仅有较少的热量收到屋顶结构中（夏天时），热空气会自动上升至屋脊处的排风孔排出，从而形成类似烟囱的抽吸效应。这种结构方式非常重要的一点是必须确保屋檐处的开孔能够确保足够的空气能够吸入到屋顶结构中。
	在确定通风空气层的厚度时，需要注意的是由于空气会随温度膨胀或者压缩，保温隔热层并不是完全平整放置的。当通风层中的空气自由循环时，若气流遇到承接梁、屋顶天窗、烟囱以及屋顶窗的时候，气流方向必然会发生改变，此时必须设置交叉木条以确保气流能够在檐口和屋脊之间自由循环（见图29）。
有支撑屋面保温隔热层	一般来说，设置有支撑屋面保温隔热层的屋顶常常并不采用通风屋顶结构。取而代之的做法是在椽木上设置支架，然后直接铺设一层预置保温隔热层。而预置保温隔热层的下表面一般具有蒸汽隔离层的功能。和全椽木保温隔热层相似，在预置保温隔热层的上面可以直接铺设屋面垫层或者其他防水层。这种屋面铺设方法可以大大缩短工期，同时由于椽木从室内仍然可见，需要作为设计元素进行处理。此时，应该选择刨平的木材或者叠合木材作为椽木使用。
内表面	为了增加室内空间的舒适性，完工后的房屋空间内表面应该设有一层既可吸收又可释放水汽的材料。一般砌体墙结构本身就具有这样的功能，而对于屋顶结构则通常采用铺设石膏板的方式，该方法非常便于现场施工。其不足之处在于当房屋结构发生微小

提示：

通风层的截面尺寸应该按照以下规定确定：当屋面倾斜角度大于10°时，檐口处的截面面积不应小于屋面面积的2‰，同时不小于200cm²，屋脊处的截面面积不应小于屋面面积的5‰。通风层中的净高度（空气层厚度）不应小于2cm；当屋面倾斜角度小于10°时，两侧屋檐处的通风层截面面积不应小于屋面面积的2‰，而通风层的净高度不应小于5cm。具体的规定可以从国家相关规范中查阅。

屋椽间保温隔热层

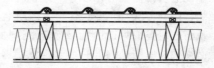

屋面覆层
屋面木条
屋面交叉木条
屋面垫层
通气层
保温隔热层
蒸汽隔离层

屋椽间保温隔热层（另设内部保温隔热层）

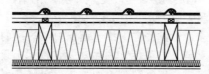

屋面覆层
屋面木条
屋面交叉木条
屋面垫层
通气层
保温隔热层
蒸汽隔离层

全屋椽保温隔热层

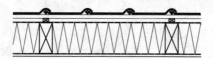

屋面覆层
屋面木条
屋面交叉木条
屋面垫层
保温隔热层
蒸汽隔离层

屋椽上保温隔热层

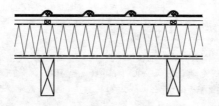

屋面覆层
屋面木条
屋面交叉木条
屋面垫层
通气层
保温隔热层
蒸汽隔离层
支架

图32：
坡屋顶的不同结构形式

图33：
屋面排水

位移时就可能引起石膏板的开裂。另一种常见的方法是铺设带波纹的企口板，企口板由于相互之间可以发生移动，所以不太容易开裂。但是，在选择内墙表面材料的时候很多情况下是出于外观的考虑。

P43

屋面排水

坡屋顶屋面雨水经过屋面和屋谷流向屋檐处的天沟，然后经天沟、落水管将雨水导入到排水系统中。如果房屋的排水系统和公共排水系统相连，则称该排水系统为<u>联合排水系统</u>；如果房屋的排水系统直接将雨水排向地下，则称该排水系统为<u>独立排水系统</u>。

排水构件尺寸确定

在确定排水构件的尺寸之前需要首先查明房屋所在地的降水量情况。

降雨量根据降雨强度计算，其中降雨强度的单位为升每秒每公顷[比如 $r=300l/(s\cdot ha)$]。而径流系数则需要根据屋顶倾斜角度、屋面性质以及屋面的平面面积确定。

屋顶的排水量则可以根据以下公式计算：

屋顶排水量 ar（单位：l/s）＝径流系数×
降雨面积 a（单位：m^2）×设计降雨强度 r

根据以上公式的计算结果查询相应的表格就可以选择合适的落水管尺寸（见表4，表5）。

天沟

排水天沟安装在檐口处高度可调整的天沟托座上。对于木结构屋顶，每榀屋椽之间设置一道天沟，天沟托座的间距与屋顶结构形式有关，但一般不应超过90cm。天沟位于天沟托座上，在天沟安装的时候需要避免天沟向外倾斜，也就是避免天沟靠近建筑的一侧高于朝外的一侧。否则，天沟中可能会有溢出的雨水直接流出建筑屋顶。天

表 4：
径流系数（摘自德国标准化学会《DIN 1986-2/ISO 1438》规范中表格 16）

表面类型	径流系数
不渗水屋面	1.0
屋面倾斜角度小于 3°	0.8
砾石瓦屋面	0.5
绿化屋面	0.3
粗放型绿化屋面（$d>10cm$）	0.5

表 5：
降雨强度为 300l/（s·ha）时不同屋面最小坡度允许采用落水管排水的最大屋面面积（摘自德国标准化学会《DIN 1986-2/ISO 1438》规范中表格 17）

屋面倾斜角度	屋面排水量（单位：l/s）	降雨面积 径流系数1.0（单位：m^2）	降雨面积 径流系数0.8（单位：m^2）	降雨面积 径流系数0.5（单位：m^2）
5°	0.7	24	30	48
6°	1.2	40	49	79
7°	1.8	60	75	120
8°	2.6	86	107	171
10°	4.7	156	195	312
12°	7.6	253	317	507
12.5°	8.5	283	353	565
15°	13.8	459	574	918
20°	29.6	986	1233	1972

表 6：
落水管和天沟设计（此处为 PVC 材料）（摘自德国标准化学会《DIN 18461/ISO 1438》规范中表格 2）

最大降雨强度为 300l/（s·ha）时对应的降雨面积（单位：m^2）	屋面排水量（单位：l/s）	落水管直径（名义尺寸）（单位：mm）	天沟（名义值）（mm）
20	0.6	50	80
37	1.1	63	80
57	1.7	70	100
97	2.9	90	125
170	5.1	100	150
243	7.3	125	180
483	14.5	150	250

必须向落水管所在位置倾斜，且坡度不应小于2%。由于金属天沟会随着温度的变化发生收缩或者伸长，所以每段金属天沟的长度不应大于15m。每段天沟之间采用连接件相连，端部天沟的外侧需要进行封闭处理。

落水管

落水管与天沟一端设有弯管，而排水天沟中留有预制孔以便于落水管的安装。落水管相互之间采用端部的卡槽和防水材料连接。建筑上预留的托座将落水管和建筑连接在一起，托座可以用销钉和螺钉校紧。落水管与建筑的间距不应小于20mm，以防止当落水管破损时潮气或雨水对建筑产生损害。

天沟和落水管既可以是圆形也可以是方形，其材料选择也很多。但在选择的时候需要特别注意避免不同材料之间发生相互影响，包括产生接触腐蚀或者发生膨胀挤压等。比如，铜制落水管和铜制天沟之间只能采用覆有铜层的铁托架和夹具进行连接。对于铝制天沟，则建议采用镀锌夹具或者铝制夹具。对于锌制或者镀锌钢管，则可以采用镀锌托架和夹具。对于PVC天沟，则可以采用镀锌托架或者覆有塑料层的托架（见表6）。

对天沟进行绿化的方式是可取的，可以在天沟处向外放置植物，从而增大天沟清洗的间隔时间。在一些特殊的位置可以布置可加热的天沟，比如在屋顶一些升起构件的根部位置（见P133提示），可加热天沟以确保冰雪天气下天沟正常排水。

内部天沟

有些天沟布置在楼板之上，而不是悬挂在檐口处（见图34），这种天沟称为"内部天沟"（见图33中左图）。当不打算或者不允许用悬挑屋顶的时候，经常采用内部天沟的做法。内部天沟可以采用安全天沟，这种天沟是在普通金属或者塑料天沟外侧另包裹一层防水材料，这样即使在天沟发生渗漏的情况下也不会对建筑造成损害。另外，内部天沟中还可以设置溢水排泄孔以应对某些特殊情况。

> 提示：
> 地方政府部分会提供不同地区在强降雨和弱降雨情况下的降雨量，内部天沟应该根据相关值的计算结果选择安全雨量大于100%的天沟。

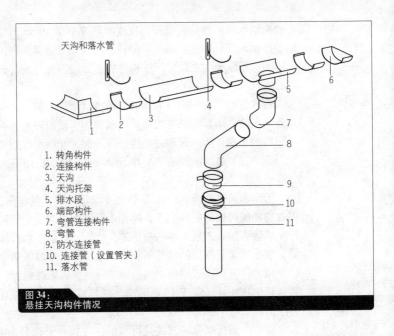

1. 转角构件
2. 连接构件
3. 天沟
4. 天沟托架
5. 排水段
6. 端部构件
7. 弯管连接构件
8. 弯管
9. 防水连接管
10. 连接管（设置管夹）
11. 落水管

图34：
悬挂天沟构件情况

P46

表现方法

拟建建筑以平面图、立面图、剖面图以及详图的方式来表现。在进行平面图绘制的时候，我们选择某一平面之上的假想平面（通常选择楼板或者某特殊楼层平面之上1~1.5m的水平面）进行绘图，也就是只有在该平面以下的结构构件才能在平面图中得以表现。所以，一般采用平面图很难非常详细地对屋顶结构进行说明。

提示：

数量单位的选择对于施工图而言非常重要。建筑和结构构件的尺寸单位一般为m（在适当的小尺寸地方，可以采用cm），尺寸链可以采用m或者cm绘制。结构构件截面尺寸一般采用cm进行标注，另外钢结构构件一般采用mm进行标注。

屋椽平面图

由于以上的原因，在坡屋顶设计时一般采用了屋椽平面图表现方法（见图35）。屋椽平面图采用顶视图的方式表现出屋顶木结构中的所有木结构构件（或者其他相应屋顶类型的相关构件），但并不显示屋顶覆层、防水层以及内部覆层等其他内容。该图是施工人员在现场施工的指南。为了标明屋顶结构（也称为屋顶桁架）的位置，在屋椽平面图中需要标出房屋中的最高楼层以及下部墙体的位置，然后根据屋顶结构和最高楼层的相对尺寸确定屋顶结构的位置。图中，采用实线表示可见的结构构件，虚线表示被遮挡的结构构件。如果支座、立柱、隅撑等结构构件被其上方的椽木遮挡，那么这些构件应该采用虚线进行表示。虽然屋面木条和屋面交叉木条都属于屋面木结构，但在屋椽平面图中并不对它们进行表达。其原因在于屋面木条和屋面交叉木条并不是由屋架施工人员进行安装，而是由屋面铺设人员随后安装的。在屋椽平面图中，不同的结构构件采用不同的编号进行分类，编号的形式一般是带有圆圈的数字，并采用线或者箭头指向结构构件。图中设有图示对不同结构构件的截面情况（宽度×高度）以及构件的材料等级等信息进行说明。

另外，在图中还需要对构件的安装和连接进行相应的说明。结构构件的截面尺寸由建筑师、结构工程师或者结构设计师确定。

除屋椽平面图之外，在建筑资料中还需要对以下结构层进行绘制和说明（见图36）：屋脊节点、屋檐、屋顶边缘做法、屋顶防水防渗层、屋面开洞以及所有的特殊处理方式（见表7）。

提示：

木结构构件的尺寸一般采用标注截面的方法进行表示。若矩形截面梁的尺寸为10cm×12cm，则表示称为10/12，念做"十乘十二"。圆形截面则采用标注直径的方法进行表示，钢结构构件采用产品型号或者截面型号的方式进行表示（比如，HEA 120、U100等）。构件的长度和定位则标注在结构设计图中。

提示：

关于结构设计图的深入讲解可以参考本套丛书中的《工程制图》一书，贝尔特·比勒费尔德和伊莎贝拉·斯奇巴编著，中国建筑工业出版社2010年2月出版，征订号：18811。

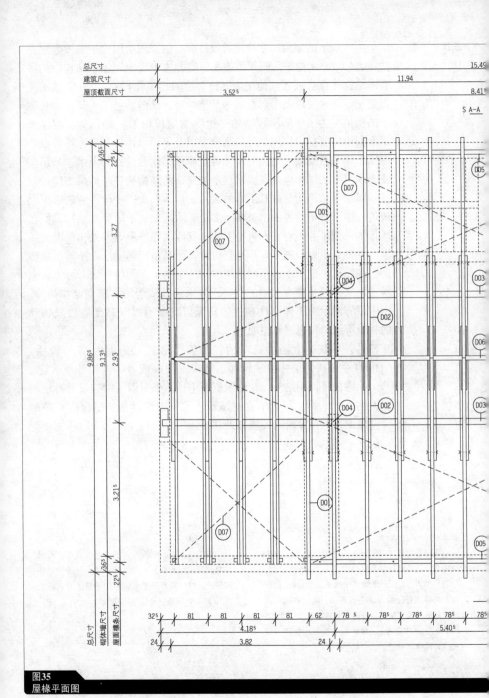

图35
屋椽平面图

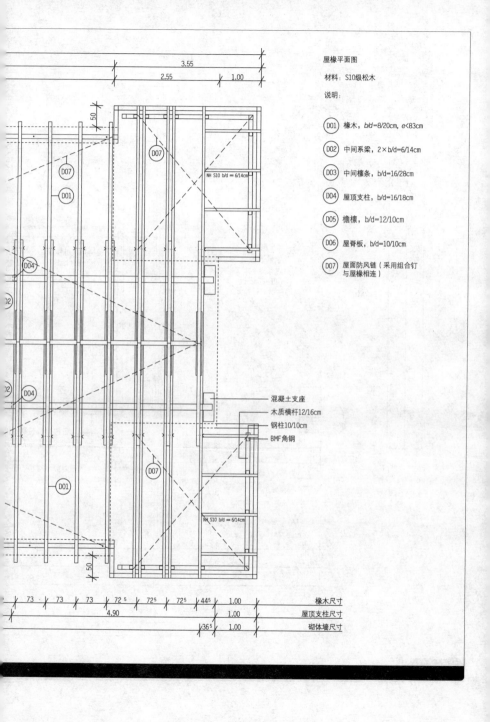

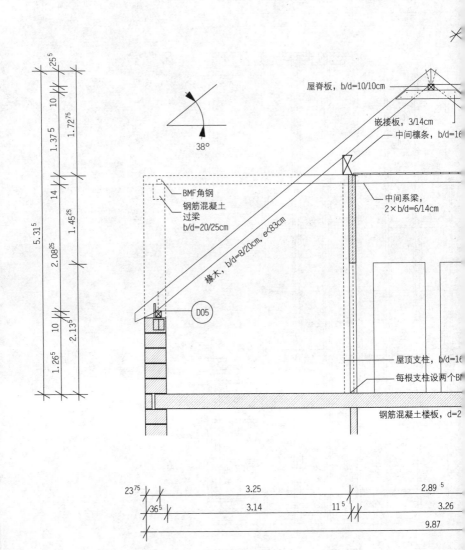

图36
屋檐剖面图

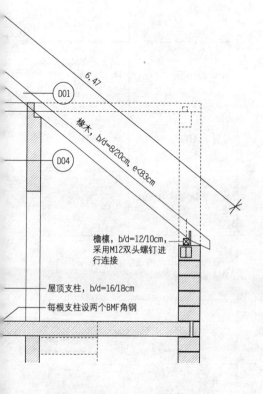

A-A剖面

材料：S10级松木

说明：

- D01 椽木，$b/d=8/20cm$，$e<83cm$
- D02 中间系梁，$2×b/d=6/14cm$
- D03 中间檩条，$b/d=16/28cm$
- D04 屋顶支柱，$b/d=16/18cm$
- D05 檐檩，$b/d=12/10cm$
- D06 屋脊板，$b/d=10/10cm$

表7：坡屋顶构件汇总表

结构构件	图示	提示	普通屋顶常用尺寸（单位：cm）
屋檐板	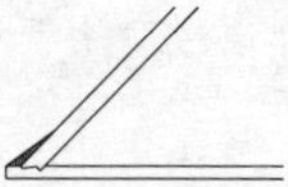	用于对椽屋架	8/12 ~ 10/22
系梁	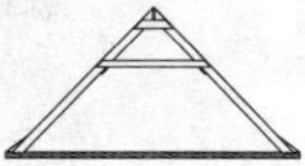	用于对椽屋架	12/12 ~ 14/14
脊檩	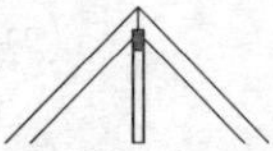	放置于墙体或立柱之上	14/16 ~ 16/22
檐檩	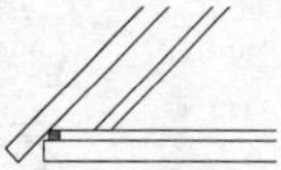	放置于顶棚或者外墙之上	10/10 ~ 14/16
中间系梁	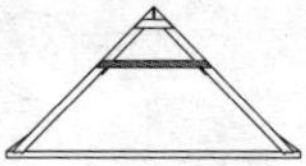	一般为成对出现的连系梁形式	8/14 ~ 10/20
轭缘	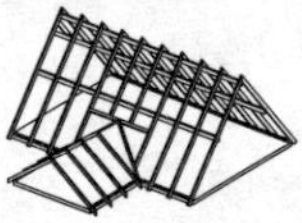	边缘板件	8/14 ~ 10/22
角钢夹板	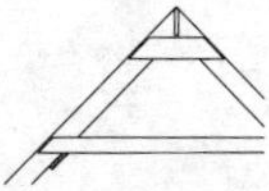	设于系杆之上	
隅撑	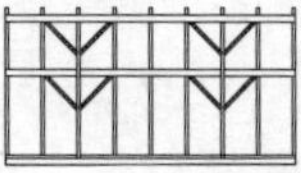	支撑于立柱之上，沿纵向布置	10/10 ~ 10/12
中间檩条	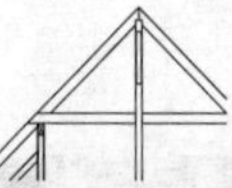	设于椽木下方	12/20 ~ 14/20

表7（续）：
坡屋顶构件汇总表

结构构件	图示	提示	普通屋顶常用尺寸（单位：cm）
立柱	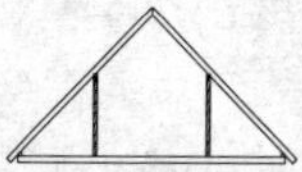	用于支撑檩条	12/12 ~ 14/14
导引构件	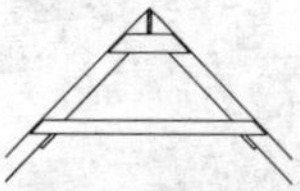	便于屋架安装	厚度不小于 22mm
椽木	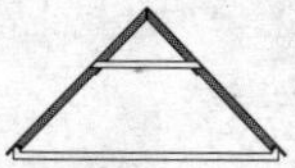	支撑屋面覆层	8/14 ~ 8/22
支撑	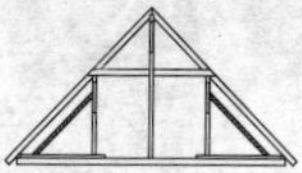	起横向加强作用	14/16
边缘构件	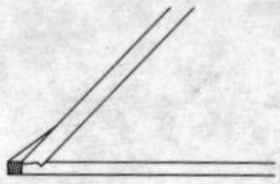	用于对椽屋架	长度 20cm
承接梁	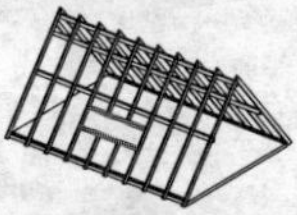	用于屋面开洞	8/14 ~ 8/22
屋面斜链支撑	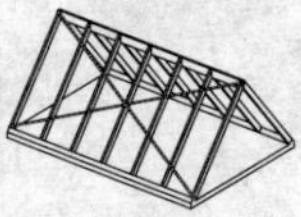	起纵向加强作用	采用扁钢
连系梁	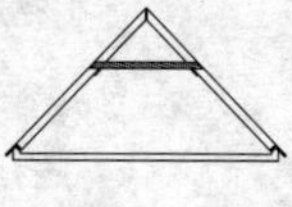	水平加强构件，一般成对出现	6/14 ~ 8/16

123

平屋顶

基础知识

平屋顶是指屋面倾斜角度小于5°的屋顶，不过平屋顶也可以用在坡度较小（不大于25°）的坡屋顶中，但此时需要确保屋面材料不会发生滑移。平屋顶的最外层为屋面覆层（见图37），一般具有耐候性或者防水性。平屋顶可以设计成为上人屋面或者上车屋面，打个比方，屋顶可以设计成为楼台、停车场或者放置通风设备。屋面需要进行精心设计，因为从邻近建筑上看，屋面就是整个建筑的第五立面，所以屋面的外观非常重要。和坡屋顶相同，平屋顶需要满足多种不同的功能：抵抗外界的湿气和降水、隔热、保温以及防风等；将屋顶自身的重量、承受的风荷载以及其承受的雪荷载、停车荷载等安全地传递到其下部结构之上。在屋顶结构所承受的荷载中，风吸力是一个比较重要的荷载，因为屋面所采用的大多是轻质的防水层，所以必须确保在风吸力的作用下不会发生防水层被掀起或者移动的情况。

平屋顶可以设计得和外墙面平齐或者悬挑出外墙。屋顶边缘升起的墙体称为女儿墙，屋顶的保温隔热层和防水层在女儿墙处进行收边。平屋顶一般采用建筑内部排水的方式，所以从屋顶边缘到落水口之间应该具有一定的坡度，该坡度不得小于2%。

屋面结构层

平屋顶具有多种不同功能的分层，而不同层之间必须相互匹配。不同层的铺设顺序需要确保屋顶结构的保温隔热和隔声的功能（见图38）。

提示：
不同的建筑类型具有不同的保温隔热和隔声标准，其处理方式需要得到相关建筑部门的许可。

屋顶必须确保结构层周围的水分（湿气以及凝结的水汽）能够有效地排出，所采用的不同材料应该能够相容。

屋面垫层　　屋面垫层设置在屋面防水层下方，可以采用屋顶结构的受力层（比如混凝土板）、木板或者保温隔热层的表面作为屋面垫层。非常重要的一点是确保屋面防水层和屋面垫层的材料能够相容，避免发生诸如膨胀收缩裂缝等现象。如果屋面铺设的是预制材料，那么在接缝处必须铺设宽度不小于20cm的分离式条带。

粘结层　　粘结层的铺设是增强屋面结构层之间的粘结力。粘结层可以采用表面处理剂，也可以采用预先铺设沥青或者沥青乳浊液的方式。粘结层可以用粉刷、轧紧或者喷射的方式铺设到洁净的屋面垫层上。

找平层　　当屋面结构层出现了不平整或者粗糙的现象时，可能需要设置找平层。找平层可以弥补结构层的误差，形成一个新的平滑表面。找平层可以采用屋顶沥青板、玻璃板、塑料棉网或者泡沫塑胶垫。找平层的铺设一般采用自由放置或者点粘结的方式。

隔离层　　隔离层的铺设是为了确保具有不同膨胀性能的相邻屋面结构层之间能够协调地发生相互运动，或者发生共同移动。当相邻的屋面结构层出现化学成分不相容时，也可以采用隔离层。隔离层可以采用与找平层相同的材料。

蒸汽隔离层　　蒸汽隔离层的铺设是为了控制屋顶空间内湿气的扩散（见"坡屋顶"一章中"面层铺设方法"部分内容）。蒸汽隔离层并不是防水层，仅能起到阻碍水汽扩散的作用，其抑制因子表示的是能够通过蒸汽隔离层扩散的水分量。蒸汽隔离层可以采用沥青屋面层、塑料隔离膜、屋面橡胶膜或者复合金属薄膜等，铺设时可以采用自由放置或者点粘结的方式。在隔离层相互连接的地方必须做到完全粘结。蒸汽隔离层必须延伸到屋面保温隔热层边缘处并进行牢固封口。为了达到室内的气密性时也可以采用蒸汽隔离层。

保温隔热层　　铺设保温隔热层的目的在于保护房屋在冬天不会损失过多的热量，同时阻止房屋在夏天吸收过多的热量。多孔聚苯乙烯板（EPS）、轧制聚苯乙烯泡沫板（XPS）、轧制聚氨酯泡沫板（PUR）、矿物纤维隔热材料（MF）、泡沫玻璃（FG）、木纤维隔热材料、轧制沥青矿物填充物均可以作为保温隔热材料使用。如果保温隔热层铺设在防水层上方，那么必须确保保温隔热层上方的所有覆层均能让水汽扩散，避免湿气聚集在保温隔热层中。保温隔热层可以设计成具有斜面的形式，以便形成屋面排水坡度。不同形式的保温隔热层可以从生产商处

> **提示：**
> 所有的屋面都必须具有不小于2%的朝向排水口的屋面坡度。如果屋面结构层本身并不具有坡度，那么可以采用设置找坡层（在屋面结构层上另外铺设一层）的方式或者铺设倾斜保温隔热层的方式来形成坡度。在确定具体采用哪种方式的时候，除去造价的因素，需要重点考虑的是屋面结构是否可以承受找坡层的重量以及倾斜保温隔热层是否能够满足建筑高度的限制要求。

> **注释：**
> 如果采用铺设保温隔热楔形块的方式进行屋面找坡，那么矩形楔形块相互之间的交线与最速下降线形成了45°的夹角。也就是说，在楔形块相互之间形成了坡度小于屋面坡度的类似屋谷线的交线。为了确保该交线的坡度不小于2%，此时屋面的坡度必须不小于3%。

购买，这种屋面保温隔热层被称为倾斜保温隔热层，而单块保温隔热材料被称为保温隔热楔形块。

铺设隔热片时应该确保隔热片相互之间具有一定的间隙，同时相互连接应尽可能牢固，保温隔热材料与隔热片的下表面相互粘结。隔热片的相互连接方法应该遵守生产商的产品说明。

蒸汽压力补偿层　　蒸汽压力补偿层的目的在于将水蒸气所产生的压力均匀地传递到屋顶防水材料上。蒸汽压力补偿层可以选用与蒸汽隔离层相同的材料，包括沥青屋面层、塑料隔离膜、屋面橡胶膜或者复合金属薄膜等，铺设时可以采用自由放置或者点粘结的方式。

屋面防水层　　屋面防水层由屋面防水材料形成。屋面防水层是一个具有防水功能的封闭区域，可以由沥青屋面层、塑料或者橡胶膜或者防水涂料铺设而成。如果采用沥青屋面层，则至少应该铺设两层，铺设时相邻沥青块之间搭接宽度不得小于80mm，并且将全部搭接面积进行粘结。如果采用塑料膜或者橡胶膜，则可以只铺设一层。全部面积需进行胶粘，搭接宽度不得小于40mm。屋面防水层铺设时必须确保覆盖全部屋顶面积。如果防水层下方的屋面层存在开口或者孔洞（比如下方为隔热片或者木结构框架），那么在防水层下方需要增设一层隔离层。屋面防水层的厚度不得小于1.5mm，对于旧屋顶则不得小于2mm。

图37：
平屋顶覆层

过滤层

  铺设过滤层（保护层）的目的在于保护屋面防水层免受外界力的影响。穿孔PVC、橡胶以及带粒塑料板、轧制聚苯乙烯泡沫排水板均可以作为过滤层材料使用。

屋面保护层

  现代金属屋面中采用了金属薄板作为单层防水层，金属屋面采用金属板带与其下方的屋面层进行粘结或者点粘结。金属屋面的常见形式为浅色金属屋面，以减小阳光照射下屋面的吸热膨胀。由于金属屋面的材料形式，这种屋面足以抵挡紫外线的照射，所以其上方不再需要另外的保护层。这种屋面形式可以减轻屋面的自重，同时也相应减小了屋顶结构的自重。当屋面不能有效抵挡紫外线照射、风吸力或者机械荷载时，需要对屋面增设保护层。根据沥青层所承受的荷载，可以设置轻型屋面保护层。比如，在沥青层上方铺砂。将碾碎的板岩撒到冷却的沥青聚合物中后也可以覆盖到沥青层上作为轻型屋面保护层。

  砂砾屋面保护层被称为重型屋面保护层，其砂砾层厚度不得小于50mm。砂砾层的重量能够避免未固定的屋面覆层被风吸力掀开。如果采用的是砂砾或者碎石，必须确保颗粒具有足够的尺寸以避免被风刮走。当屋面承受大风的时候，建议对屋面保护层进行覆盖。

可上人
屋面层

  可上人屋面层可供人行走，也可以起到屋面保护层的作用。若采用了可上人屋面层，那么保温隔热层就应该选择抗压缩材料。屋面防水层必须得到适当的保护，以避免机械荷载的影响。

  当可上人屋面的楼板进行了密封处理或者屋面覆层封闭时，屋面需要设有不小于1%的坡度。排水装置可以设在防水层中，此时的防水层需要设耐水层。楼板必须选择防冻楼板，相应的连接也应该采用伸缩接头，以应对温度涨缩效应。在距离屋顶边缘的区域，应该有足

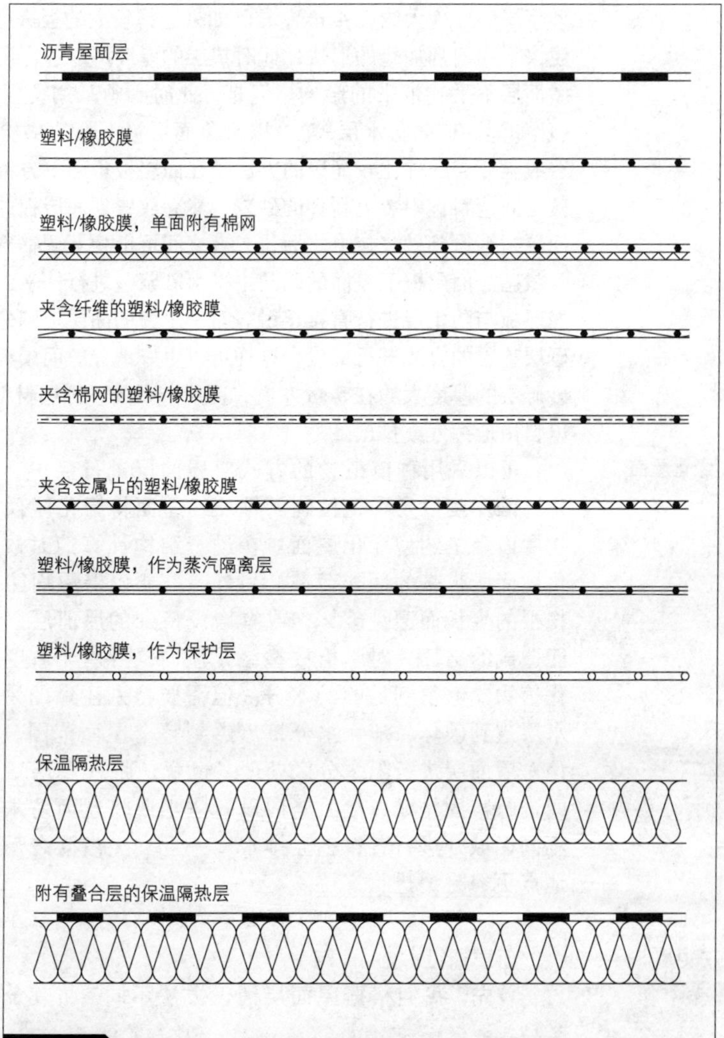

沥青屋面层

塑料/橡胶膜

塑料/橡胶膜，单面附有棉网

夹含纤维的塑料/橡胶膜

夹含棉网的塑料/橡胶膜

夹含金属片的塑料/橡胶膜

塑料/橡胶膜，作为蒸汽隔离层

塑料/橡胶膜，作为保护层

保温隔热层

附有叠合层的保温隔热层

图38：
平屋顶屋面层介绍——根据 Zentralverband des Deutschen Dachdeckerhandwerks；Flat roof guidelines 所介绍的屋面结构层

够的间隔，以避免对该处卷起的屋面防水层造成破坏。可上人屋面层可以选择铺设在垫有绒垫或者排水层上的砂浆层中的小板形式，此时砂浆层的厚度大约在4cm左右。如果选择较大的板进行铺设的时候则建议采用角部加垫的形式，此时垫层的厚度可以进行调整以克服下部屋面层不平整带来的影响。但是，此时必须采用抗压缩防水层材料（比如泡沫玻璃防水层等），以避免面层板下部的垫层对防水层挤压造成破坏。一种比较简单的方法是在面层板角部下方铺设砂浆包，也就是向塑料包中装上新拌的砂浆，然后放置到面层板的角部下方。当面层板铺设到砂浆包上之后，二者之间可以相互调整到紧密接触的平整状态。面层板下表面的凹凸不平由砂浆包进行抹平，同时也避免了面层板与防水层进行直接接触。对于更大的面层板则建议采用铺设砂砾层垫层的方式进行铺设，这样可以更好地分散面层板的重量。此时砂砾层的厚度大约在5cm左右，而所选择的砂砾层材料要确保水分可以自由地挥发或排出。

绿化屋面　　可以采用种植植物的方式对屋面层进行保护。此时，针对绿化层的厚度以及种植植物的性质，屋面将绿化屋顶分为粗放型绿化屋面和集约型绿化屋面。在进行结构计算的时候，必须考虑绿化屋面所带来的额外荷载。另外，需要确保屋面防水层不会因植物根的生长而遭受破坏，或者增设特殊的屋面层以防止植物根可能造成的破坏。对于粗放型绿化屋面，可用无坡度屋面以确保绿化植物能够得到足够的水分。但需要特别注意的是，如果屋面防水层遭到破坏，那么水分将渗透到整个屋面范围。所以，此时应该对屋面层进行隔水分区处理，也就是将（多层）防水层相互之间进行牢固粘贴，然后竖向设置隔水壁将屋面防水层分隔成若干不同区域，进行各自独立排水。一旦发生渗漏之后，可以对渗漏地点<u>更容易地进行定位</u>。

P60　　　　**面层铺设方法**

　　首先，我们根据屋面层铺设方法不同，将其分成主要的三种类型。

不通风屋面　　不通风屋面（之前称为"保暖屋面"，见图39）的屋面防水层设在屋面外层，此时屋面保温隔热层则位于屋顶内部的"保暖区域"。典型的不通风屋面在施工过程中一般会在屋面承重层（包括钢筋混凝土楼板、钢结构楼板以及木结构楼板）上预先铺设一层覆层。然后在预铺层上在进行找平层和蒸汽隔离层的铺设。保温隔热层则采用的是耐磨斜面板，并且确保其朝向屋面排水孔或者天沟的方向坡度不小于2%（3%更佳）。在保温隔热层上方铺设蒸汽压力补偿层，然后再进行防

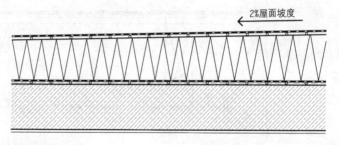

屋面防水层（单层）
蒸汽压力补偿层
保温隔热层（单层斜面板）
蒸汽隔离层
找平层
预铺层
钢筋混凝土楼板

图 39：
不通风屋面

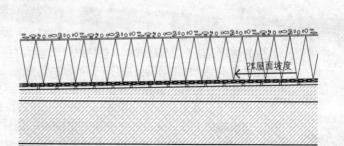

屋面防护层，砂砾层，厚度不小于5cm，颗粒直径16~32mm
过滤层
保温隔热层，耐久性防水材料
屋面防水层（3层）
找平层
找坡层
钢筋混凝土楼板

图 40：
倒置屋面

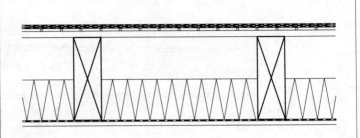

屋面防水层（3层）
找平层
平板支撑，22mm
空气流通层，高度不低于15cm
保温隔热层
椽木
蒸汽隔离层
石膏板

图 41：通风屋面

水层的铺设，防水层可以选择单层的也可以选择多层的。根据选择材料性质的不同，应该对屋面进行不同的保护措施。

倒置屋面

屋面铺设的第二种方式是倒置屋面（见图40），其安装过程也被称为屋面保温隔热膜安装（IRMA）。此处，屋面保温隔热层铺设于屋面防水层的上部，所以要求必须采用抗水性材料。对于这种屋面结构，屋面应设置坡度不小于2%的找坡层。找坡层上方设置找平层，然后铺设屋面防水层。而保温隔热层则可以采用平面板材的形式。保温隔热层的上方设置过滤层，防止屋面保护层中的颗粒被冲刷至保温隔热材料中。

通风屋面

第三种屋面铺设方法是通风屋面（以前也称为"冷屋面"，见图41），这种铺设方法在木结构屋顶中使用得较多。屋面承重结构下方覆有石膏板或者硬纸板，而蒸汽隔离层设置于石膏板与椽木的下表面之间。保温隔热层设置于椽木之间，可以采用单层形式。保温隔热层上方和椽木之间应至少保留15cm的净空，以确保有足够的空间供空气流通。

椽木的上方设有平板支撑，材料可以选择硬纸板、舌榫板或者类似的材料。然后依次铺设找平层和屋面防水层，在有需要的地方可以铺设屋面保护层。

P63
屋面升起结构

防水板

为了防止雨水的溅起或者聚集造成其向屋面内部渗透，必须在屋面边缘处设置屋面升起结构，将屋面防水层向上升起一段高度。另外，屋面升起结构还包括屋面装饰、烟囱或者设备室等。对于屋顶上的门窗等，也需要采用相同的措施。对于屋面倾斜角度不到5°的平屋面，屋面防水材料需要升起到屋面边缘和屋面升起结构15cm以上，但并不需要到屋面升起结构的顶面，其顶面可以选择砂砾保护层。如果屋面坡度大于5°，那么屋面防水材料在屋面升起结构上的升起高度不应小于10cm。

门槛

阳台或者屋顶露台门槛处的防水是一个难点。如果将此处的屋面防水材料从屋面层上升15cm，那么必然将在室内外之间形成一个台阶。大多数情况下，室内外的地面保持了表面标高相同但室外结构标高低于室内结构标高的做法，以消除室外防水层升起所造成的影响，不过这种做法额外地增加了一些补偿措施。为了更容易出入屋顶空间，可以将门槛的高度降低到5cm。但这种做法必须确保在多雨雪的天气下，水分不会从屋面防水层后方渗透到屋面结构中。如果门外附近没有设置排水孔，那么需要在门槛处设置格栅或者天沟进行排水。对于"无障碍建筑"，比如一些公共建筑等，要求必须选择无门槛门。此时则需要采用一些特殊的施工方法，比如设置遮雨棚防止溅水、采用与排水系统直接相连的加热天沟或者采用全粘结整体屋面结构等。

提示：
　　屋面上墙、柱等竖向构件均为屋面升起结构。当屋面防水材料铺设到屋面升起结构处时，需要特别注意避免在屋面防水材料下方发生雨水直落、飞溅或者凝结等现象。

屋面防水层和隔离层均可以采用夹紧装置进行封口，然后固定到屋面升起结构的墙体上。塑料屋面防水层可以与组合金属板进行粘结，然后对接口进行夹紧处理，并采用伸缩接头与屋面结构相连以避免发生张拉损坏。防水层升起的部分可以采用搭接在夹紧装置上的切角金属板进行覆盖保护，金属板的另一边则延伸至屋面保护层中。在金属板的覆盖过程中需要特别小心，避免其对防水层造成损伤。对于窗立面，即使防水层被人为活动所破坏的可能性更大，防水层也应该延伸到窗背面。

所有的屋面组成材料都应该满足相应的防火规范要求。比较特殊的区域是屋顶升起的部分，一般来说屋内的火不会通过屋面开洞传播到屋面升起的地方（比如上部采光圆屋顶等等）。但是此时依然应该在屋面上方保证5m的火隙，以避免可能发生的轰燃现象。

平屋面部分施工

屋面收边

屋面到墙面、竖杆之间的部分均可以作为平屋面部分进行施工，目前的做法一般是在边缘处设置女儿墙或者切边的方法进行收边（见图43）。平屋面部分施工存在一些难点，比如屋面防水层和保温隔热层与墙体中的防水层、保温隔热层之间的牢固连接。如果将屋面保温隔热层进行延伸，对整个屋面部分进行铺盖至外墙部分，则可以避免冷桥区域形成，但是这种做法导致屋顶层厚度过大。另一种做法是将保温隔热层铺设在屋顶结构层的内部，但是这种方式导致屋面保温隔热层不可能与外墙保温隔热层连接，同时这种做法也使屋面结构层处于室外较低的温度里，进而导致屋面结构层与外墙或者柱所承受的温度效应不同。温度效应的不同可能会对结构连接处进行张拉，从而导致结构或者立面中产生裂缝。这种做法一般用在为了保留已有旧建筑外观，而对其进行后增保温隔热层的铺设中。

女儿墙、升起结构部分施工

当屋面遇到竖向构件时，屋面防水层可以向上升起，其做法与屋面升起结构相似。此时，屋面防水层升起屋面的高度不得小于10cm，其参考面为屋面的最上层——金属薄板、砂砾层或者上人屋面层等。升起的竖向构件可以采用与女儿墙相同的形式，比如采用钢筋混凝土结构、砌体结构或者采用切边加边缘盖板的形式。盖板一般可以用普通矩形木板制作而成，能平整地铺设在下方的保温隔热材料上。切边部分往往制作成包裹屋面边缘并与立面部分重叠的形式。如果采用女

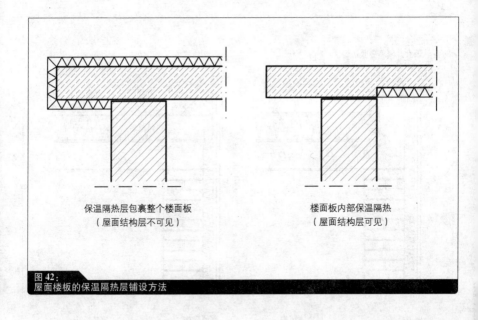

保温隔热层包裹整个楼面板　　　　　楼面板内部保温隔热
（屋面结构层不可见）　　　　　　　（屋面结构层可见）

图42：
屋面楼板的保温隔热层铺设方法

儿墙的形式，那么一般会采用金属薄板（也可以采用预制石材或者预制混凝土块）来对屋面边缘进行包裹。根据建筑的高度不同，其所受到的风荷载也存在较大的不同，建议相应采用的女儿墙尺寸也不同。在女儿墙立面墙体以上2cm左右的地方应该设置滴水檐，以防止雨水聚集到金属板的后方（见表8）。

设女儿墙屋面收边做法 切边式屋面收边做法

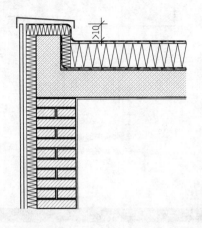
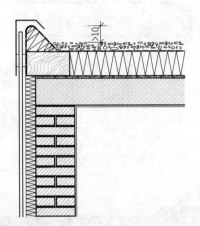

屋顶结构	墙体结构	屋顶结构	墙体结构
单层屋面防水层	墙面砖	砂砾层	墙面砖
蒸汽压力补偿层	通风层	单层屋面防水层	通风层
保温隔热层	硅酸钙保温隔热层	蒸汽压力补偿层	硅酸钙保温隔热层
蒸汽隔离层	砌体墙	保温隔热层	砌体墙
找平层		蒸汽隔离层	
预铺层		找平层	
钢筋混凝土楼板		预铺层	
		钢筋混凝土楼板	

图 43:
女儿墙以及切边屋面收边做法

表 8:
女儿墙盖板与立面墙搭接长度要求

建筑高度	搭接长度
<8m	5cm
8~20m	8cm
>20m	10cm

屋面排水

平屋顶一般会采用内部排水装置,也就是说落水管从建筑内部将屋面水排放到地面以下。每一个屋顶至少需要设置一个排水孔和一个紧急排水孔(见图45),而落水管的尺寸的确定方法与坡屋顶相同(见"坡屋顶"一章"屋面排水"部分内容)。(见图46)

屋面坡度　　屋面施工时必须确保朝向排水孔的坡度不小于2%,也就是说屋面不可能是完全水平的。屋面的坡度可以由找坡层或者倾斜保温隔热层形成(见"保温隔热层"部分内容),用以防止发生小范围的积水。对于上人屋面,屋面的顶层和保温隔热层均需要考虑排水。屋面集水沟或者排水孔应该设置在屋面高度最低的位置,确保雨水能够自由地排出。若采用集水沟排水,那么集水沟距屋面升起构件以及屋面连接的位置应不小于30cm,此时可以在上面设置溢水格栅,避免排水孔中进入杂物。

屋面排水孔　　对于屋面集水沟,必须含有防水连接环对金属板之间进行连接,或者设有翼缘与两侧的防水层连接,并进行沥青封边。平屋顶中所使用的集水沟既可以采用垂直进水孔,也可以采用倾斜进水孔。但一般情况下,更倾向于采用垂直进水孔,因为垂直进水孔能够将雨水直接排到落水管中,而且在发生渗漏的情况下能够更快地进行定位。当进水孔涉及过多的区域或者当雨水无法直接排出的时候(比如对于无柱空间),可以采用倾斜进水孔与落水管连接进行排水。

位于暴风雪区域的建筑,建议采用可加热排水孔以防止发生冻结现象,确保在冬天的时候也能够有效地排水。

紧急排水装置　　为了在一个落水孔堵塞(比如被树叶遮挡)的情况下,防止屋面发生积水现象,必须至少设置一个紧急排水装置(见图44)。紧急排水装置可以是设置在屋顶边缘(女儿墙)内侧的排水管或者排水沟,其应该布置在屋面的较低位置。必须选用防水材料,并对周边进行保温隔热。排水管或者排水沟需要伸出屋面足够的长度,以防止其排出的雨水飞溅到墙面上。由于紧急排水装置并不是永久性排水设置,所以并不需要与地面排水系统连接。

图44:
紧急排水装置

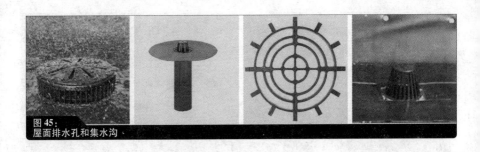

图 45:
屋面排水孔和集水沟

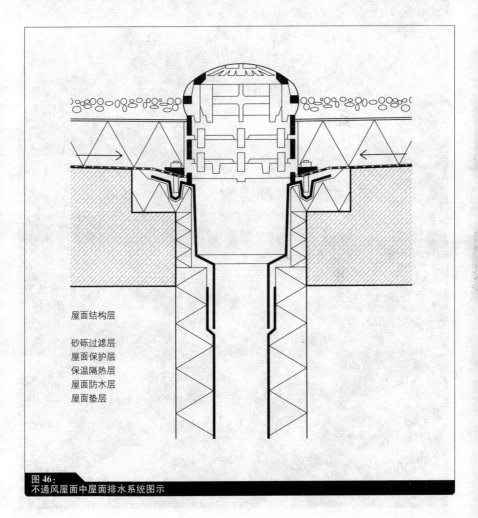

屋面结构层

砂砾过滤层
屋面保护层
保温隔热层
屋面防水层
屋面垫层

图 46:
不通风屋面中屋面排水系统图示

结语

前文主要对屋顶建筑的不同形式、材料以及设计方法进行了介绍。本书仅仅涉及简单形式的屋顶,但实际情况中并没有这样标准形式的屋顶——而是存在不同可能的组合形式。设计者需要对屋顶的结构层铺设、连接方式、封口做法以及抗渗做法等问题进行考虑和设计。在进行以上问题的细节设计时必须对以下几个问题进行充分考虑:

形式:
— 所选择的屋顶形式与建筑规模以及平面设计是否相适应?
— 屋顶是否具有第二个逃生通道(比如窗户等)?
— 所选择的屋顶以及屋顶窗(若有)形式是否符合相应的建筑规范?
— 所选的屋顶材料是否与建筑整体外观以及周边建筑相协调?
— 屋顶空间是否具有足够的采光和通风条件?

铺设:
— 建筑的外表面是否设有足够的保温隔热层(包括山墙、屋顶窗、屋顶表面、屋顶上部结构以及上升管道等)?
— 屋顶结构中的保温隔热层是否与周边结构(外墙、阳台等)中的保温隔热层进行了可靠连接?
— 屋顶面层是否能够可靠地防雨?
— 屋面结构构件周围空气中的水分是否能够有效地排出?
— 是否所有屋面上的雨水都能够畅通无阻地流向排水系统?
— 屋顶面层是否能够可靠地防风?
— 是否能有效地阻止水汽的凝结(比如,渗透点附近)?
— 是否有效地预防了空气中的水分向屋面层(特别是屋面保温隔热层)中渗透?

对于屋顶结构,除了常规设计内容之外还需要对一些其他的内容进行设计。如今,一些常规的建筑可以被设计成令人兴奋的新形式,但这些设计从整体上仍然体现出了建筑所应该遵守的基本规定。在进行屋顶设计的时候,设计者不仅需要遵守建筑所应该符合的规范和原则,更重要的是首先在头脑中应该形成屋顶建筑的基本概念,屋顶的样式则是下一步需要考虑的内容。

附录

相关规范

荷载作用

E DIN 1052	Entwurf, Berechnung und Bemessung von Holzbauwerken – Allgemeine Bemessungsregeln und Bemessungsregeln für den Hochbau, Berlin, Beuth-Verlag 2000 (Design, calculation and dimensioning for timber buildings)
E DIN 1055-1	Einwirkungen auf Tragwerke – Teil 1: Wichte und Flächenlasten von Baustoffen, Bauteilen und Lagerstoffen, Berlin, Beuth-Verlag 2000 (Effect on load-bearing systems, part 1, weights and area loads for materials, parts and stored material)
E DIN 1055-3	Einwirkungen auf Tragwerke – Teil 3: Eigen- und Nutzlasten für Hochbauten, Berlin, Beuth-Verlag 2000 (Work on load-bearing systems, part 3, own weight and imposed loads for building)
E DIN 1055-4	Einwirkungen auf Tragwerke – Teil 4: Windlasten, Berlin, Beuth-Verlag 2001 2001 (Effect on load-bearing systems, part 4, wind loads)
E DIN 1055-5	Einwirkungen auf Tragwerke – Teil 5: Schnee- und Eislasten, Berlin, Beuth-Verlag 2000 (Effect on load-bearing systems, part 5, snow and ice loads)

封边做法

DIN 18195	Bauwerksabdichtungen, Teile 1-6 und 8-10, Ausgaben 8/83 bis 12/86, Berlin, Beuth-Verlag 1983/1986 (Sealing buildings)

保温隔热

DIN 4108	Beiblatt 2, Wärmeschutz und Energie-Einsparung in Gebäuden. Wärmebrücken. Planungs- und Ausführungsbeispiele (1998-08) (Supplementary sheet 2, heat insulation and energy saving in buildings, heat bridges, examples of planning and execution)
DIN 4108-2	Wärmeschutz und Energie-Einsparung in Gebäuden. Mindestanforderungen an den Wärmeschutz (2001-03) (Heat insulation and energy saving in buildings, minimum demands)

DIN 4108-3	Wärmeschutz und Energie-Einsparung in Gebäuden. Klimabedingter Feuchteschutz, Anforderungen, Berechnungsverfahren und Hinweise für die Planung und Ausführung (2001-07) (Heat insulation and energy saving in buildings, climate-related damp protection, requirements, calculation procedures and hints for planning and execution)
DIN V 4108-4	Wärmeschutz und Energie-Einsparung in Gebäuden. Wärme- und feuchteschutztechnische Bemessungswerte (2002-02) (Heat insulation and energy saving in buildings, heat and damp protection technical dimension values)
DIN 4108-7	Wärmeschutz und Energie-Einsparung in Gebäuden. Luftdichtheit von Gebäuden. Anforderungen, Planungs- und Ausführungsempfehlungen sowie -beispiele (2001-08) (Heat insulation and energy saving in buildings, air-tightness of buildings, requirements, planning and execution recommendations and examples)
SN EN ISO 10211-1	Wärmebrücken im Hochbau – Berechnung der Wärmeströme und Oberflächentemperaturen – Teil 1: Allgemeine Verfahren (ISO 10211-1:1995), 1995 (Heat bridges – calculating heat currents and surface temperatures, part 1, general procedures)
SN EN ISO 10211-2	Wärmebrücken im Hochbau - Berechnung der Wärmeströme und Oberflächentemperaturen - Teil 2: Linienförmige Wärmebrücken (ISO 10211-2:2001), 2001 (Heat bridges – calculating heat currents and surface temperatures, part 2, linear heat bridges)
屋面排水	
DIN 18460	Regenfallleitungen ausserhalb von Gebäuden und Dachrinnen (Rainfall drainage outside buildings and gutters)
DIN EN 612	Hängedachrinnen, Regenfallrohre ausserhalb von Gebäuden und Zubehörteile aus Metall, Berlin, Beuth-Verlag, 1996 (Suspended roof gutters, drainpipes outside buildings and metal components)
SN EN 612	Hängedachrinnen mit Aussteifung der Rinnenvorderseite und Regenrohre aus Metallblech mit Nahtverbindungen, 2005 (Suspended gutters with reinforced gutter fronts and sheet metal drainpipes with seam joints)
SSIV-10SN EN 12056-3	Schwerkraftentwässerungsanlagen innerhalb von Gebäuden – Teil 3: Dachentwässerung, Planung und Bemessung, Ausgabe 2000 (Gravity drainage inside buildings, part 3, roof drainage, planning and dimensioning)

P75 参考文献

Francis D.K. Ching: *Building Construction illustrated*, 3rd edition, John Wiley & Sons, 2004

Andrea Deplazes (ed.): *Constructing Architecture*, Birkhäuser Publishers, Basel 2005

Thomas Herzog, Michael Volz, Julius Natterer, Wolfgang Winter, Roland Schweizer: *Timber Construction Manual*, Birkhäuser Publishers, Basel 2004

Ernst Neufert, Peter Neufert: *Architects' Data*, 3rd edition, Blackwell Science, UK USA Australia 2004

Eberhard Schunck, Hans Jochen Oster, Kurt Kiessl, Rainer Barthel: *Roof Construction Manual*, Birkhäuser Publishers, Basel 2003

Andrew Watts: *Modern Construction Roofs*, Springer Wien New York 2006

P76 图片来源

图3左一、图33左二（德国维特住宅）：德国维特，Tanja Zagromski

图3左二、左三，图23左二，图25左二、左三，图26右一，图37左一、左三，16页、54页图：德国多特蒙德，贝尔特·比勒费尔德

图7右一（德国奥尔登堡住宅），图33左三、右一（德国洛登住宅）：柏林阿尔弗雷德·迈斯特曼（Alfred Meistermann）

图页40（德国Naels住宅）：德国埃森，布鲁宁

图33左一（Klenke公寓楼），图44（别墅57+）：德国波鸿，Archifactory.de

本书其他图片均由作者提供。